INDIVIDUALISED MATHEMATICS

Developed by the School Math
association with the National

PROBABILITY AND STAT

CAMBRIDGE UNIVERSITY PRESS

Cambridge
London   New York   New Rochelle
Melbourne   Sydney

The School Mathematics Project

When the SMP was founded in 1961, its main objective was to devise radically new secondary school mathematics courses to reflect, more adequately than did the traditional syllabuses, the up-to-date nature and usages of mathematics.

SMP Books 1–5 form a five-year course leading to the O-level examination in SMP Mathematics. *Revised Advanced Mathematics Books 1, 2 and 3* cover the syllabus for the A-level examination in SMP Mathematics. Five shorter texts cover the material of the various sections of the A-level examination SMP Further Mathematics. There are two books for SMP Additional Mathematics at O-level. All the SMP GCE examinations are available to schools through any of the GCE examining boards.

Books A–H cover broadly the same development of mathematics as the first few books of the O-level series. Most CSE boards offer appropriate examinations. In practice, this series is being used very widely across all streams of comprehensive schools and its first seven books, together with *Books X, Y and Z*, provide a course leading to the SMP O-level examination. *SMP Cards I and II* provide an alternative treatment in card form of the mathematics in *Books A–D*. The six units of *SMP 7–13*, designed for children in that age-range, provide a course for middle schools which is also widely used in primary schools and the first two years of secondary schools. Teachers' guides accompany all these series.

The SMP has produced many other texts, and teachers are encouraged to obtain each year from the Cambridge University Press, P.O. Box 110, Cambridge CB2 3RL, the full list of SMP publications currently available. In the same way, help and advice may always be sought by teachers from the Executive Director at the SMP Office, Westfield College, Kidderpore Avenue, London NW3 7ST. The annual Reports, details of forthcoming in-service training courses and other information may be obtained from the SMP Office.

The SMP is continually evaluating old work and preparing for new. The effectiveness of the SMP's work depends, as it has always done, on the comments and reactions received from teachers and pupils in a wide variety of schools using SMP materials. Readers of the texts can, therefore, send their comments to the SMP in the knowledge that they will be taken into consideration.

The authors of the original books on whose work this series is based are named in *The School Mathematics Project: The First Ten Years*, published by the Cambridge University Press.

SMP Individualised Mathematics has been produced by a team consisting of

Judy Bonsall	G. Merlane
G. S. Howlett	L. Savins
M. K. Leach	D. R. Skinner
J. L. Lloyd	J. V. Tyson

John Lloyd led the work on the series until his death in 1977, and the final editing has been carried out by Derek Skinner. Many others have helped with advice and criticism, particularly those students who worked through the material in draft form.

Contents

Preface

SMP Individualised Mathematics is based upon the content of *SMP Books 1–5* and *Books A–G, X, Y, Z*, covering the syllabus for the O-level SMP Mathematics.

There are two main distinguishing features of the series. First, the material is presented in a programmed form and the books are thus suitable for use in individualised learning, where self-assessment and clear explanation play a major role. The carefully structured development of each topic makes the books suitable for use by students working alone with minimum tuition, in schools, technical colleges, colleges of further education and on courses organised by the National Extension College.

Secondly, instead of the spiral development of the SMP texts, *SMP Individualised Mathematics* presents the mathematics by topics. Each book, apart from the two devoted to revision, presents the work on a particular theme. Hence the books will prove useful to pupils who have missed work through absence from class, to students coming from abroad, and to pupils transferring to a different school. The style and arrangement of these books should make them very suitable for use by pupils in the sixth form who are working to improve their earlier performance at CSE or O-level. The books will also be useful for revision and consolidation.

Although written with the SMP O-level course in mind, *SMP Individualised Mathematics*, like other SMP texts, can be used to prepare for other O-level examinations based on similar syllabuses.

The titles in this series are as follows:

Computation and Graphs
Probability and Statistics
Algebra 1: Language and Structure
Algebra 2: Equations, Formulas and Graphs
Further Algebra and Computation
Matrix Algebra and Isometric Transformations
Further Matrices and Transformations
Geometry 1: Symmetry and Trigonometry
Geometry 2: Shapes and Similarity
Geometry 3: Three Dimensions
Revision 1
Revision 2

How to use this book

Each chapter begins with a list of what you should be able to do after studying the chapter. This is followed by a pre-test, which gives you some idea of what you should know before you start that particular chapter. If you have difficulty with the pre-test, you should revise the work required for it – from either the appropriate chapter of this or a companion book or an elementary text-book – before continuing with the chapter.

The teaching part of the chapter is divided into several sections, and includes a number of exercises. Other questions are asked in the text, and *you should write down the answers to all these questions and exercises in a notebook* as you go along. The start of each set of questions is marked by a white triangle on the left-hand side of the page. When you come to a triangle with a number in it (on the *right-hand* side of the page) you should check your work to that point by turning to the answers at the end of the chapter and finding the triangle with the same number (now on the *left-hand* side of the page).

The teaching part of the chapter is followed by a summary of the important results of the chapter (you may well find it helpful to copy these into a separate notebook that is kept especially for revision), and a post-test to test your understanding of the chapter as a whole. The answers to this post-test are also at the end of the chapter.

Finally (apart from the answers) there is an 'assignment'. This is another exercise covering the whole chapter, but this time there are no answers in this book. If at all possible you should have *this* exercise marked by a teacher or tutor.

1 Statistics

Objectives

This is what you should be able to do after studying this chapter.
(1) Represent statistical data by bar charts, pictograms and pie charts.
(2) Find the initial data if you are given a pie chart or a bar chart.
(3) Form a frequency table from a collection of data.
(4) Identify the mode, the median, and the mean of a collection of data.

Pre-test

1 Work these out.

(a) 6.1×3.2 (b) 16×270 (c) $\frac{2.51}{1.27}$ (correct to 3 s.f.)

(d) $\frac{360}{112} \times 53$ (correct to the nearest whole number)

2 On graph paper draw x- and y-axes for values of x from 0 to 8 and values of y from 0 to 400. Plot the points A (2, 200), B (5, 155), C (8, 375), and D (7, 80).

3 Draw a circle with radius 5 cm. Draw in one radius. Use your protractor to draw angles of 50°, 90°, 27°, 42°, 100°, 41° and 10° between successive radii, so that the circle is divided into seven parts (sectors). What is the sum of the angles?

1.1 Introduction

Elementary statistics involves collecting information or *data* (usually in a numerical form); representing this information in some sort of diagram; and then (sometimes) finding a representative, or *average*; i.e. a member that in some way is typical of the information collected.

For the purposes of this chapter the following examples will be used as collections of data. As most of them will be used several times, they are printed here for convenience.

Example 1

The method used to travel to work by 3600 workers in a factory.

Car	1300
Cycle	800
Bus	400
Train	900
Foot	200

Example 2

The vitamins, etc. (in milligrams), contained in 500 grams of cereal.

B_1	3.5 mg
B_2	5.0 mg
B_6	6.0 mg
F_1	22.5 mg
N_1	53.0 mg

Example 3

The average annual production of copper between 1950 and 1970 (in millions of tonnes).

USA	1.38
Zambia–Zaire	1.14
Chile	0.78
USSR	0.66
Canada	0.60
Peru	0.24
Others	1.20

Example 4

The number of families of different sizes, for the families of a class of children in which there were no brothers or sisters.

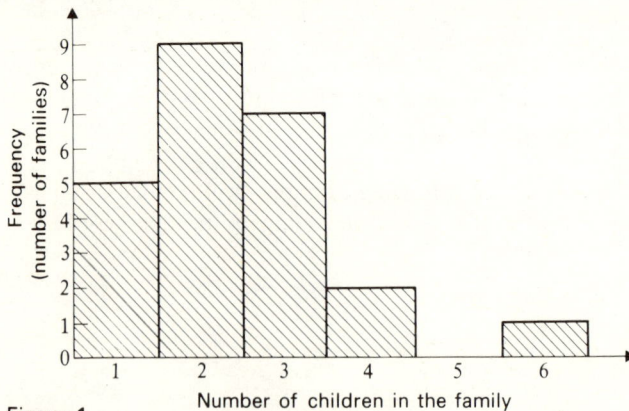

Figure 1

Example 5

The heights (in centimetres) of twenty-one girls in a class.
167 170 173 167 172 172 174
175 165 167 172 172 174 166
168 166 171 174 169 172 169

Example 6

The shoe sizes taken by twenty-nine second year pupils.
2, $2\frac{1}{2}$, 3, 3, $3\frac{1}{2}$, 4, 4, 4, 4, $4\frac{1}{2}$, $4\frac{1}{2}$, $4\frac{1}{2}$, $4\frac{1}{2}$, $4\frac{1}{2}$, $4\frac{1}{2}$, $4\frac{1}{2}$, 5, 5, 5, 5, 5, 5, 5, 5, $5\frac{1}{2}$, $5\frac{1}{2}$, 6, 6, 6, 6

Example 7

The length, in words, of the first twenty-five sentences of a novel.
9, 7, 3, 1, 7, 4, 1, 2, 5, 13, 54, 12, 10, 21, 3, 6, 5, 6, 7, 8, 13, 15, 6, 7, 7

Example 8

The marks out of 100 obtained by twenty-four people in an examination.
29, 58, 73, 82, 37, 44, 30, 82, 44, 61, 29, 44, 91, 11, 38, 58, 46, 49, 70, 48, 63, 82, 57, 58

1.2 Visual representation

The most obvious impact of statistics is visual (compare Example 4 with the other examples); and in this section we shall consider how we can represent the data of Example 1 in a diagram.

Bar charts

One method is to draw a bar chart. The number of people travelling by car (1300) is represented by a rectangle, or *bar*, 1300 units long (or tall), using a suitable scale. The other methods of transport are treated in the same way. The width of the bars doesn't matter, provided they are all of the same width: sometimes the bars are little more than thick lines.

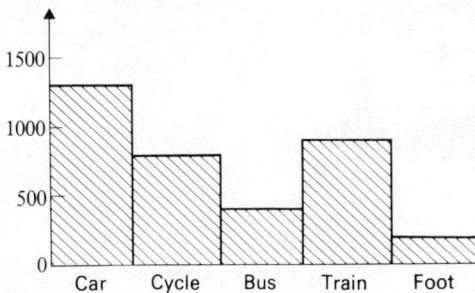

Figure 2

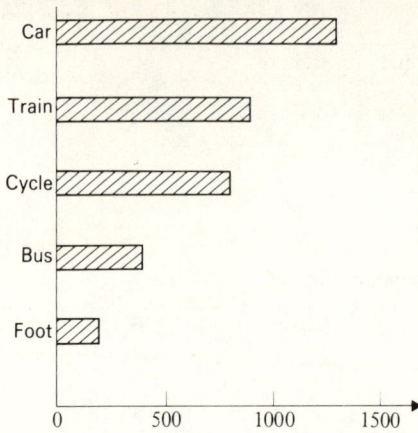

Figure 3

Figures 2 and 3 are exactly the same as far as the display of information is concerned; they are merely different layouts to suit different tastes. In either case the bars can be touching (as in Figure 2) or separated by a small gap (as in Figure 3). In examples such as this, where the 'things' (cars, trains, etc.) are not numerical, the order of the bars is not important: in Figure 2 it is the same as that given in the data; in Figure 3 the order has been rearranged to go from the smallest to the largest. But in situations such as Example 4 where the 'things' are numbers (in this case, number of children in a family), it is obviously sensible to put them in order.

Pictograms

An alternative form of a bar chart is a pictogram. This is a bar chart with the scale removed, and the bars replaced by a series of small pictures (called symbols or motifs). Each complete motif represents a standard number – in Figure 4 each picture represents 200 people using that form of transport.

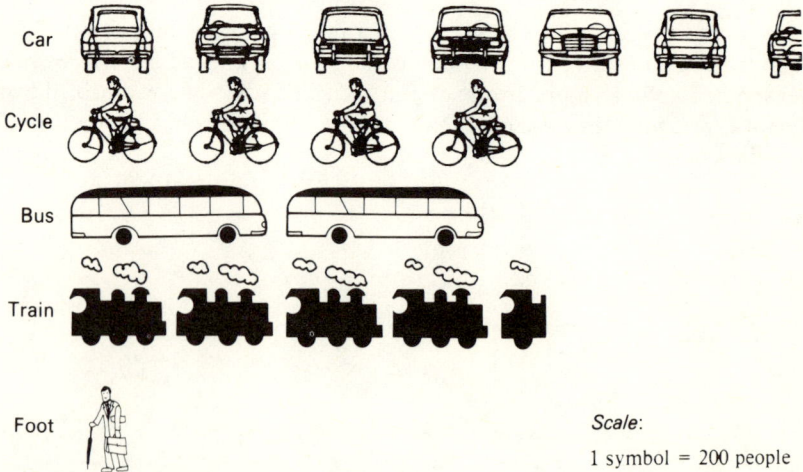

Scale:

1 symbol = 200 people

Figure 4

4

It might be better to use the same motif for all the bars; but artistic licence is allowed as this type of representation is used mostly to give a quick, dramatic, but not necessarily very accurate, picture. It is not easy to decide how much of a train, for example, to draw to represent 100 people!

Exercise A

1 Represent the information in Example 2 by (a) a bar chart; (b) a pictogram.

2 Repeat question **1** for Example 3.

3 In Example 4, Figure 1 is a *frequency diagram*, a special type of bar chart that shows, for example, that there were five families with only one child.
   (a) How many families had three children in them?
   (b) How many children were there in the largest family?
   (c) How many families are represented altogether in this bar chart?

Pie charts

A third form of visual representation takes the form of a cake or pie; with parts (sectors) marked off to represent the data. In Figure 5, the number of degrees of each sector is proportional to the number of people using that form of transport.

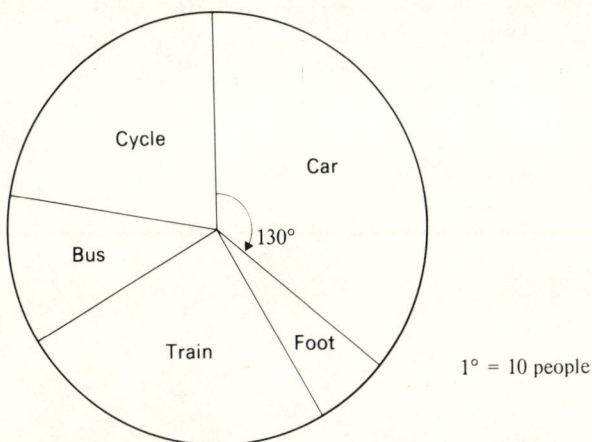

Figure 5

The angles for the pie chart are worked out by the following steps.
First, find the total number to be represented. In Example 1 this is 3600 people.
Next, work out the scale, remembering that the total number has to be represented by 360°. In Example 1, therefore,

$$360° \text{ represents } 3600 \text{ people}$$
$$1° \text{ represents } \quad 10 \text{ people } (3600 \div 360)$$

that is, each person is represented by $\frac{1}{10}°$.
Finally, work out the angles for each sector. For example, the angle for those travelling by car is $1300 \times \frac{1}{10}° = 130°$; and by cycle is $800 \times \frac{1}{10}° = 80°$; etc.
Don't forget to give your pie chart a title, and a 'scale'.

5

This is the monthly budget for a family.

Rent and rates	£84
Heating and lighting	£36
Travel	£24
Food and drink	£78
Clothing	£15
Entertainment	£33

How do we show this on a pie chart? Since the total expenditure is £270, this means that 360° has to represent £270.

$$£270 \text{ is to be represented by } 360°$$
$$£1 \text{ is to be represented by } \tfrac{360°}{270} \ (= 1.33\ldots°)$$

So the angle for 'Rent and rates' will be $84 \times 1.33\ldots° = 112°$; that for 'Heating and lighting' will be $36 \times 1.33\ldots° = 48°$; and so on. (This is an occasion when the CONSTANT key on a calculator is useful.)

▷ 1   Copy and complete the table, and then compare your answers with the corresponding angles on the pie chart in Figure 6.

Item	Cost	Angle		
Rent and rates	£84	$84 \times 1.33\ldots°$	=	112°
Heating and lighting	£36	$36 \times 1.33\ldots°$	=	48°
Travel				
Food and drink				
Clothing				
Entertainment				

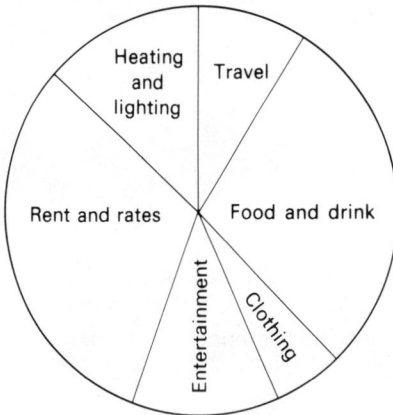

$$10° = £7.50$$

Figure 6

6

Exercise B

> **1**   Construct pie charts for Examples 2 and 3.

2   A travel agent carried out a survey of the holiday habits of students. He asked 240 of them to state the main type of accommodation they had used during the previous summer holiday. The results are illustrated in Figure 7.

(a)   Work out the scale used for the pie chart.

(b)   Measure the angle of each sector, and work out how many students were represented in each category.

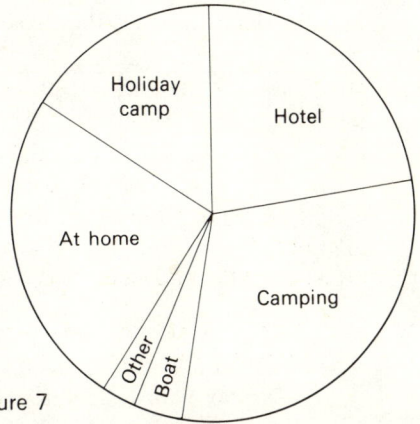

Figure 7

3   Pie charts are not always circular! It takes 300 minutes to recover the components from a scrapped car. Figure 8 shows how that time is divided between the components. Find the times taken to recover each component.

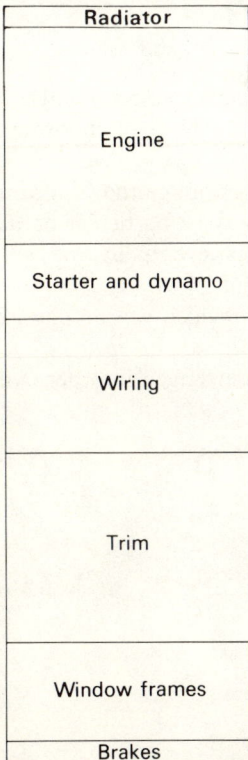

Figure 8

▷ **4** The following surveys are suggested for practical work.

(a) Find the 'most common letter', that is which letters are used the most, and which the least, in the English language. In order to do this, choose an (English) book and make a selection of, say, ten non-consecutive pages. Then make a tally table for the letters of the alphabet. (For example, the tally table for the last sentence is shown here. Note that it is usual to 'count in fives' when making a tally table. Thus there are six a's listed in the table; the tally for the fifth one is the stroke that 'crosses out' the first four.)

a	ᚍ l	f	‖	k	l	p	l	u	
b	‖	g		l	ᚍ	q		v	
c		h	‖‖	m	l	r	‖	w	
d		i		n	l	s	l	x	
e	ᚍ ‖l	j		o	‖	t	ᚍ ‖l	y	l
								z	

From your tally table, make a frequency table. (For the tally table above this would be a – 6, b – 2, c – 0, etc.)

Display your results using one or more of the methods described, and arrange the letters in order of frequency.

(b) Repeat the procedure, but use a book printed in a different language. (The order of frequency for French is given in the answers.)

▷ 5

1.3 Choosing a representative number

Parliamentary elections are concerned with choosing one person to represent a large group of people. In this section we shall examine some of the ways of choosing a single number to represent a large group of numbers.

Think about some of the groups of people to which you belong, and how you would set about selecting a representative from that group, for a particular occasion. It is a complicated business, which is often simplified so that eventually you will have to select one person from a 'short list' of three or four.

In the rest of this chapter we consider a 'short list' from which we can choose a number to represent a group of numbers.

Figure 9 illustrates the data of Example 5. The girls are arranged in order, with the shortest on the left, and the tallest on the right.

Figure 9

▷ **1** What is the height of the tallest girl?

2 What is the height of the shortest girl?

8

3 What is the height halfway between the tallest and the shortest?
$$\left(\text{i.e. } \frac{\text{tallest height} + \text{shortest height}}{2}\right)$$

4 What is the most common height?

5 What is the height of the girl in the middle?

6 What is the 'average' height if we add up all the heights and divide by the number of girls?

The best way to find the answers to some of these questions is to make a frequency table (as in question **4** of Exercise B).

Height	Tally	Frequency	Height	Tally	Frequency
165	I	1	170	I	1
166	I I	2	171	①	1
167	I I I	3	172	THHL	5
168	I	1	173	I	1
169	I I	2	174	I I I	3
			175	I	1

For example, you can now see that the most common height is 172 cm; and, counting up or down to the middle, or eleventh, girl (the tally with a circle round it), the answer to question **5** is 171 cm.

Write down the answers to all the questions.

Exercise C

1 Answer the following questions for Example 7. (Use questions **1** to **6** above as a guide.)
(a) What is the length of the longest sentence?
(b) What is the length of the shortest sentence?
(c) What is the length halfway between the longest and the shortest?
(d) What is the most common length?
(e) What is the length of the middle sentence?
(f) What is the 'average' length, if all the lengths are added and divided by the number of sentences?

2 (a) Which two of the six representative numbers obtained in question **1** would you definitely reject as not being representative of the group as a whole?
(b) Would you expect question **1**(c) to give a representative number generally? (Consider Example 6: the size halfway between 2 and 6 is 4, but how does this compare with the data?)

1.4 The mode

You will have realised that it is not easy to know what method to use in choosing a representative number from a group of numbers. We shall now look at three possible choices in more detail.

1 Which was the most popular form of transport in Example 1?

2 Which country produced the most copper in Example 3?

3 Which was the commonest height in Example 5?

4 Which was the commonest size of family in Example 4?

In these examples you have found one possible representative name or number by finding the most popular, or the commonest, name or number; i.e. the one that occurs most frequently.

This value is called the *mode*. Note that the mode is the *thing* that occurs most often, not the number of times it occurs.

Exercise D

1 Which is the mode in each of Examples 2 and 7?
Is the mode necessarily a number?

2 Which is the mode in each of Examples 6 and 8?
Is it possible to have more than one mode?

3 What is the modal record in any current 'Top Ten'?
How is the order for the 'Top Ten' worked out?

1.5 The median

The mode is easy to find, but it does not take into account any values other than its own; and for a small number of items it is often not very 'typical'. In Example 8, a mark of 82 is not really very typical of that class! (But if we had the results of, say, 1000 candidates then the mode is quite likely to be typical; and in Example 6 a shoe size of $4\frac{1}{2}$ is also typical of that group.)

An alternative representative number that does, to some extent, take into account all the other numbers in the data is the *median*.

1 What is the height of the girl in the middle of Figure 9?

2 What is the size of shoe taken by the middle person in Example 6?

3 From your tally table for Example 7 (see question **1** of Exercise C) find the length of the 'middle sentence'.

4 If the families in Example 4 are arranged in size, how many children are there in the 'middle family'?

In each of these examples you have found a representative number for each group by finding the 'middle number' *when the group has been rearranged in order of size.* This value is called the *median*.

Why is it not sensible to try to find medians for Examples 1–3?

5 Find the median for each of these groups of data.
(a) 2, 4, 1, 7, 5
(b) 3, 6, 6, 5, 7, 1, 8
(c) 1, 2, 3, 3, 4, 5
(d) 1, 2, 3, 4, 5, 6

10

What problems did you have with (c) and (d)?

In (c) there are two middle numbers, but as they are both 3 it is reasonable to say that the median is 3. In (d) there are two middle numbers, 3 and 4. Hence any number between 3 and 4 *would* do for the median, but in practice we take the halfway number, i.e. $3\frac{1}{2}$ in this example. (The median is sometimes defined as 'a number such that half the members of the data are equal to or greater than the median, and half the members are equal to or less than the median'.)

Exercise E

1 Arrange each set of numbers below in order of size, and find the median.
(a) 7, 6, 2, 1, 5, 8, 3
(b) 3, 1, 6, 3, 2, 4
(c) 4, 8, 9, 1, 2, 5, 7, 4, 3, 10
(d) 21, 25, 31, 28, 22, 29, 30, 24

2 To what extent is the median affected by the extreme values in the data?

3 Find the median mark for Example 8.

1.6 The mean

We have seen that the mode and the median are representative numbers that are fairly easily obtained, but they are not affected very much by the high and low values of the data. We are now going to consider a third representative, a number that is significantly affected by every member of the data.

A teacher gives the same test to two different classes in order to compare their performances. The marks obtained (out of a maximum of 10) are as shown.

Class XX	4, 5, 6, 9, 10, 8, 7, 6, 4, 4, 8, 5, 4, 8, 9, 5, 4, 6, 10, 8
Class YY	3, 8, 5, 4, 5, 5, 6, 7, 7, 4, 6, 6, 7, 4, 6, 7, 5, 4, 3, 7, 5, 7, 8, 3

The teacher starts by making frequency tables for each class, and then drawing the corresponding frequency diagrams as in Figure 10.

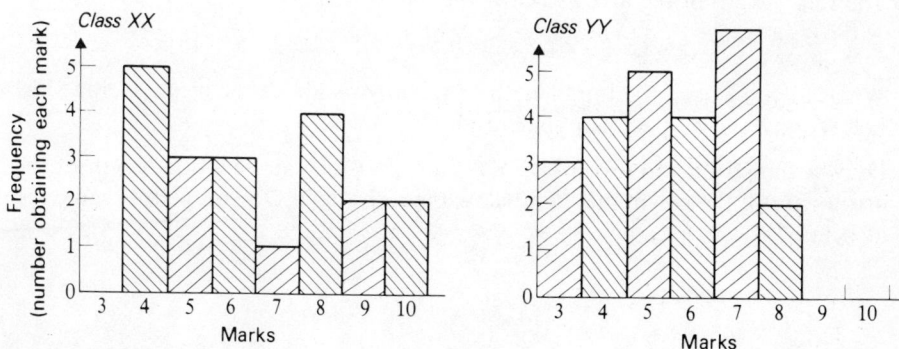

Figure 10

11

▷ 1   How many pupils were there in each class?

2   What is the mode in each class? Are they good representatives? Do they give a fair comparison between the two classes?

3   What is the median mark for each class? Are these good representatives? Do these give a fair comparison between the classes?

4   When the teacher adds up the marks for each class, the totals are 130 for Class XX, and 132 for Class YY. Why can't we use these totals as a good basis for comparison?

5   Divide 130 by 20 (the number of pupils in Class XX) and 132 by 24 (the number in Class YY). Does this give us a better basis for comparison? ◁12

You will probably agree that the last results (6.5 for Class XX and 5.5 for Class YY) are the best comparisons; and quite likely you are already familiar with them as 'the average' in each case.

The mathematical name for this type of average is *mean average, arithmetic mean*, or simply the *mean*. Thus the mean for Class XX is 6.5 and for Class YY is 5.5.

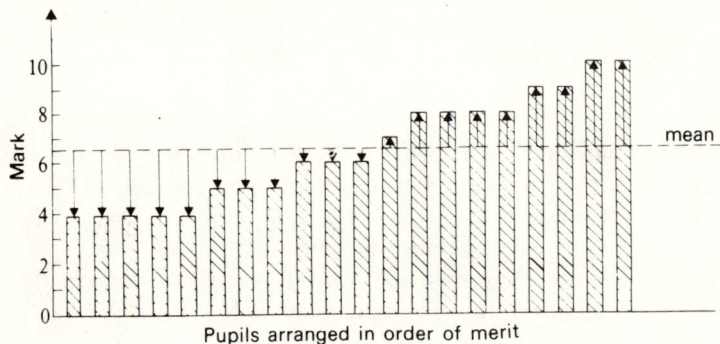

Figure 11

Figure 11 shows how Class XX's marks compare with the mean of 6.5.

▷ 6   Did the same number of pupils obtain marks above the mean as below the mean?

7   What do the arrows above and below the mean average line show?

8   The total length of the arrows below the line is:

$$5 \times 2.5 + 3 \times 1.5 + 3 \times 0.5 = 12.5 + 4.5 + 1.5 = 18.5$$

What do you expect the total length of the arrows above the mean average line to be? Work it out and explain your result.

9   Draw a similar diagram for Class YY's marks. Calculate the total length of the arrows above the mean line, and below the mean line. Do you get the same results as before? ◁13

Exercise F

1   Calculate the mean number of children per family for Example 4.

2   Ten packets of sweets are found to contain 18, 16, 20, 19, 17, 16, 20, 19, 20, 18 sweets respectively. What is the mean number of sweets per packet?

3   Calculate the arithmetic mean of 2, 4, 6, 8, 3, 7.
    Use your answer to write down the arithmetic means of these numbers.
    (a)   102, 104, 106, 108, 103, 107
    (b)   72, 74, 76, 78, 73, 77
    (c)   1442, 1444, 1446, 1448, 1443, 1447

4   (a)   The arithmetic mean of twelve numbers is 7. What is their sum?
    (b)   The mean age of a group of girls is 12 years 1 month. If there are twenty-four girls in the group, what is the sum of their ages?

5   (a)   After nine completed innings, a batsman's average was 17 exactly. In his next innings he was out for 7 runs. What is his average now?
    (b)   How many runs must he score in his next complete innings to bring his average up to 19?

6   (a)   Look back at your answers to question **1** of Exercise C and write down the mean value of the sentence length in Example 7.
    (b)   Find the mean average mark in Example 8, correct to 1 decimal place.
    (c)   Should the mean always be a whole number?

7   Using your answers to questions **1–6** in Section 1.3, construct a bar chart for Example 5 similar to Figure 11, and mark on it the mode, median and mean. Comment on each as a suitable 'representative number'.

Summary

(1)   *Visual presentation*
Statistical information may be represented visually by means of a bar chart, pictogram, or pie chart.
    A public opinion poll was taken just before a General Election to show how people intended to vote.

Conservative	Labour	Liberal	Others
41%	35%	21%	3%

In the pie chart in Figure 12 the angle for the Conservatives is worked out as follows:
$$41\% \text{ of } 360° = \tfrac{41}{100} \times 360° = 148° \text{ approximately}$$

(2)   *Frequency table and frequency diagram*
The First Division football matches on one bank holiday resulted in the following scores.

    0–1   2–1   0–3   1–1   0–2   3–2   1–2   1–1   2–1   1–1

The frequency table shows, for example, how often a particular number of goals was

13

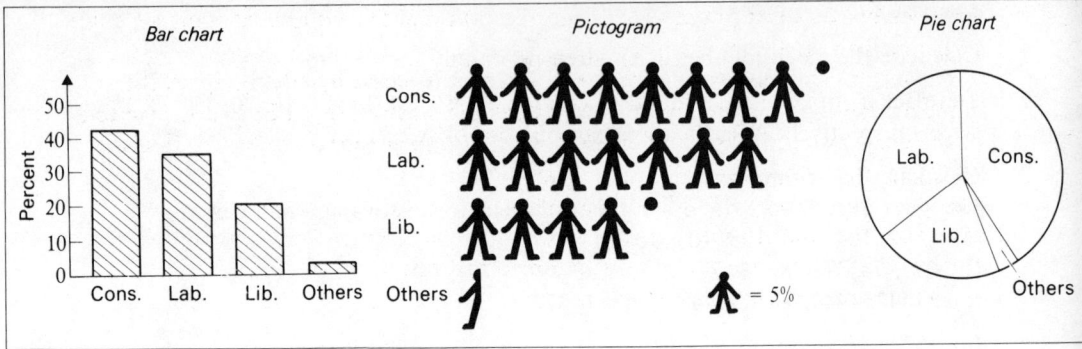

Figure 12

scored by one team; the frequency diagram (Figure 13) is a special type of bar chart to display this information.

No. of goals scored in a match by a team	Frequency
0	3
1	10
2	5
3	2

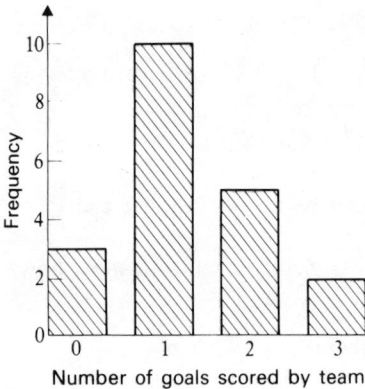

Figure 13

(3) *Representative numbers, or averages*

There are three numbers that are often used as 'averages' or typical members of a group of numbers.

The mode is the commonest, most popular or most frequently occurring number. In Figure 13 it is '1 goal'.

The median is the number in the middle (or halfway between the two middle numbers) when the figures are arranged in order. In Figure 13 there are twenty

14

teams involved: when put in order both the 10th and the 11th teams scored 1 goal, so the median is '1 goal'.

The mean (arithmetic mean, or mean average) is found by adding up all the numbers and then dividing by how many numbers there are. In Figure 13, the total of the goals scored is 26, there are 20 scores, and so the mean is $\frac{26}{20} = 1.3$ goals.

(4) *Degree of accuracy*

It is conventional to give the mode and the median to the same degree of accuracy as in the data, and the mean to one more significant figure, as in the examples above. (Cricket averages are an exception to this rule!)

Post-test

1   In a survey on the time taken to repair electrical goods, the following information was obtained.

Repairs carried out in less than one week	
Automatic washing machines	71%
Twin-tub washing machines	74%
Dishwashers	69%
Cookers	60%
Refrigerators	54%

(a)   Illustrate this information by a bar chart.

(b)   Why is it not possible to show this information by a pie chart?

2   The pie chart in Figure 14 shows where each £1 of the rates levied by a County Council was spent.

(a)   Measure the angles of the pie chart, and then express the three items as percentages of the whole.

(b)   Illustrate the information in the form of a pictogram, using a symbol of a 5p piece as the unit.

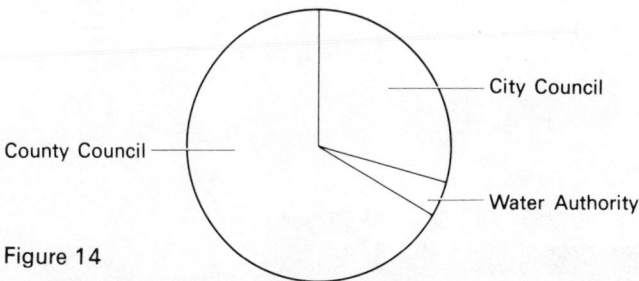

Figure 14

3   A shopkeeper sold fireworks at various prices. The prices, in pence, of the first 60 that he sold one day are given below.

```
1 1 6 3 2   3 6 9 2 4   2 4 6 1 2   4 9 5 4 1
2 2 1 4 5   3 3 9 9 6   1 9 6 4 1   5 2 4 6 2
3 2 1 1 9   6 2 1 1 2   6 1 5 3 1   1 9 1 2 1
```

15

(a) Construct the frequency table, and draw a frequency diagram.

(b) Find the modal, median, and mean prices.

Assignment

1 For a certain school, the number of passes at each grade of an examination are given in the table.

A	B	C	D	E	Unclassified
11	30	19	8	13	13

(a) Show this information on a pie chart.

(b) Express these results as percentages.

(c) These are the national figures for that particular examination.

A	B	C	D	E	Unclassified
10%	35%	20%	$7\frac{1}{2}$%	$12\frac{1}{2}$%	15%

Complete the bar chart in Figure 15 to compare the school's results with the national figures.

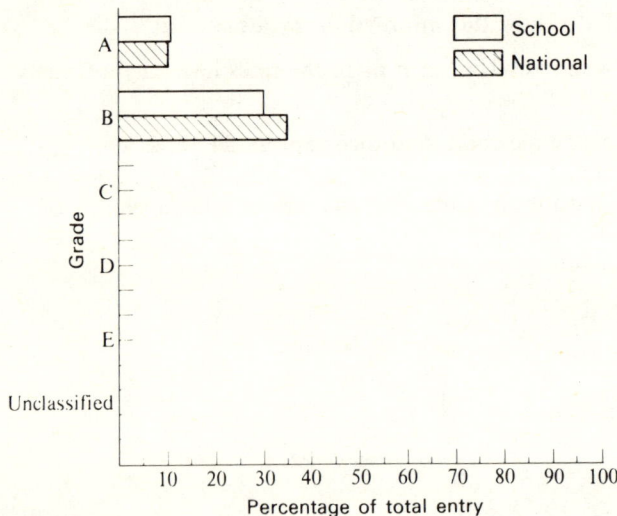

Figure 15

2 The pie chart in Figure 16 shows the value of the exports of Dodoland for 1978 under three headings: I = food, etc., II = raw materials, III = manufactured goods. If the total exports had a value of £2 400 000, calculate the values of the three separate groups.

16

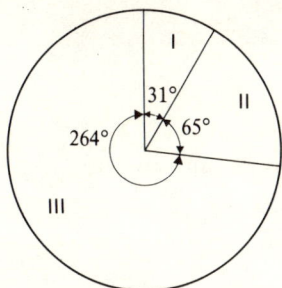

Figure 16

3 The table gives the number of hours of sunshine per week for the four weeks of August in 1977–9 for a seaside resort. If you were publishing the holiday guide for 1980, would you quote as the average amount of sunshine per week (a) the mode; (b) the median; (c) the mean for 1979; (d) the mean for 1977–9 inclusive?

Week	1	2	3	4
1977	51	72	44	65
1978	49	58	64	74
1979	70	44	62	53

Answers

Pre-test

1 (a) 19.52   (b) 4320   (c) 1.98   (d) 170

2 See Figure A. Suitable scales would be 1 cm to 1 unit on the x-axis and 1 cm to 50 units on the y-axis, but other scales could be used, of course, if you are not given one.

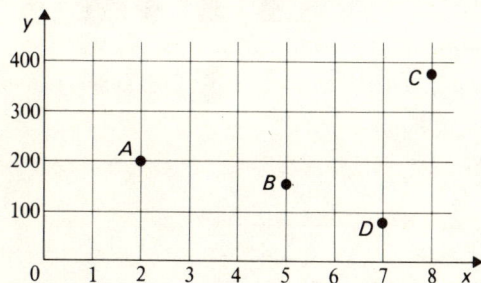

Figure A

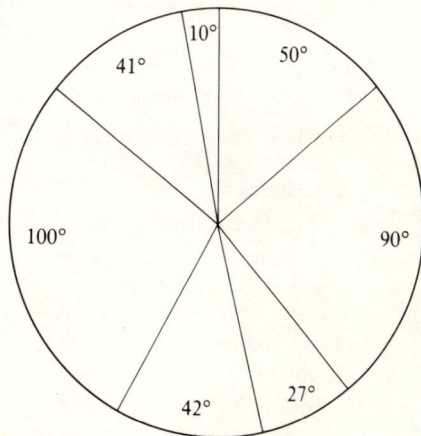

Figure B

3 See Figure B. The sum of the angles is 360°.

1.2 Visual representation

Exercise A

1   (a) See Figure C   (b) See Figure D
2   (a) See Figure E   (b) See Figure F
3   (a) 7 families   (b) 6 children   (c) $5+10+7+2+0+1 = 25$ families

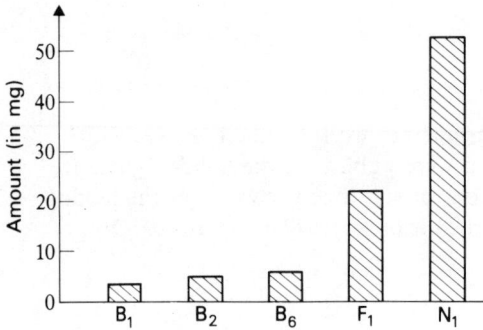

Figure C

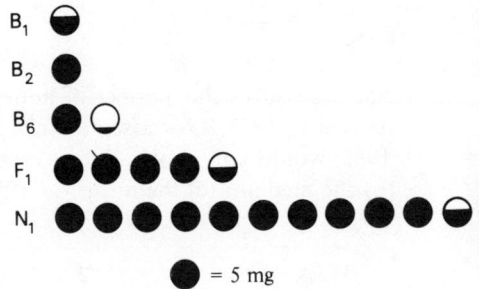

$\bullet$ = 5 mg

Figure D

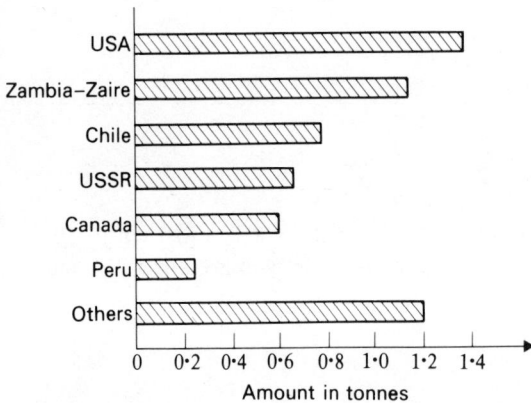

Figure E

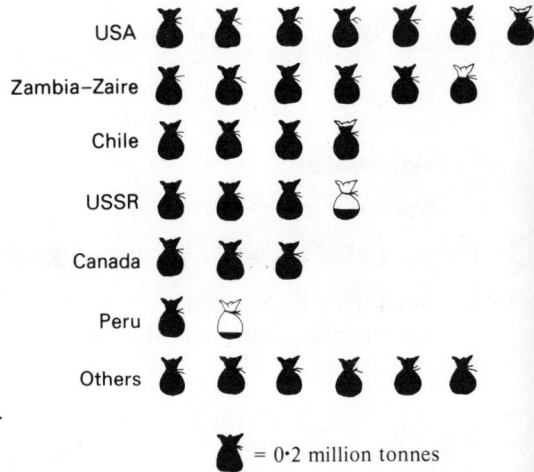

= 0·2 million tonnes

Figure F

Pie charts

1

Item	Cost	Angle		
Travel	£24	$24 \times 1.33\ldots°$	=	32°
Food and drink	£78	$78 \times 1.33\ldots°$	=	104°
Clothing	£15	$15 \times 1.33\ldots°$	=	20°
Entertainment	£33	$33 \times 1.33\ldots°$	=	44°

When you have worked out the angles, check that they add up to 360° (or maybe 359° or 361° if you have approximated the angle representing one unit).

18

Exercise B

> 1   *Example* 2
See Figure G. The angles are as follows:

B_1	B_2	B_6	F_1	N_1
14°	20°	24°	90°	212°

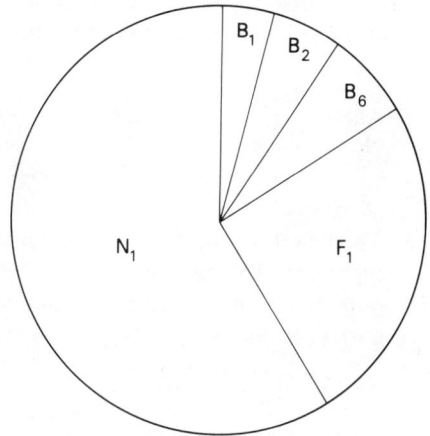

Total = 90 mg (or scale: 4° = 1 mg)

Figure G

Note that the *position* around the pie doesn't matter.

Example 3
See Figure H. The angles (to the nearest degree) are as follows:

USA	Zambia–Zaire	Chile	USSR	Canada	Peru	Others
83°	68°	47°	40°	36°	14°	72°

Total = 6·0 million tonnes
(or scale: 6° = 0·1 million tonnes

Figure H

2 (a) 360° represents 240 students, and so 1° represents $\frac{240}{360} = \frac{2}{3}$ of a student;
(or 3° represents 2 students; $1\frac{1}{2}$° represents 1 student).

(b)

	At home	Holiday camp	Hotel	Camping	Boat	Other
Angle	90°	57°	81°	108°	15°	9°
No. of students	60	38	54	72	10	6

3 The length of the complete rectangle is 10 cm or 100 mm. Therefore 100 mm represents 300 minutes; or 1 mm = 3 minutes. Measurements to the nearest millimetre are as accurate as can be expected, and so the following times will be correct to the nearest minute or two.

Radiator	Engine	Starter and dynamo	Wiring	Trim	Window frames	Brakes
9	84	30	54	75	39	9

5▷ 4 (a) The overall order of frequency in the English language is:

ETOAN IRSHD LCWUM FYGPB VKXQJZ

If you persevered with a large sample, your results should be close to this; but for a small sample the results could differ appreciably.

(b) The order of frequency in French is:

ENASR IUTOL DCMPV FBGXH QYZJKW

1.3 Choosing a representative number

6▷ 1 175 cm **2** 165 cm **3** $\frac{175+165}{2} = 170$ cm **4** 172 cm **5** 171 cm

6 The sum of all the heights is 3575 cm. $3575 \div 21 = 170.2$ cm (correct to 1 decimal place).

Exercise C

7▷ 1 The frequency table is as follows:

Sentence length (in words)	1	2	3	4	5	6	7	8	9	10	12	13	15	21	54
Frequency	2	1	2	1	2	3	5	1	1	1	1	2	1	1	1

(a) 54 words   (b) 1 word   (c) $\frac{1}{2}(54+1) = 27\frac{1}{2}$ words   (d) 7 words   (e) The middle sentence is the 13th, and after they have been put in order of length this has 7 words.   (f) 232 words $\div$ 25 sentences gives an 'average' of $9\frac{1}{4}$ words per sentence.

2   (a)   (a) and (b) are fairly obviously not typical.

(b)   In Example 6 (shoe sizes), (c) could be a reasonable representative. In this example (sentence length) it certainly isn't!

   (If the frequency diagram for the data is fairly symmetrical, and the largest and the smallest are not 'way out', then (c) can often be a reasonable representative.)

1.4   The mode

1   Car   **2** USA   **3** 172 cm   **4** Families with 2 children

Exercise D

1   In Example 2 the most common 'vitamin' is N_1; in Example 7 a sentence with 7 words is the most common. The mode is not always a number.

2   In Example 6 sizes of $4\frac{1}{2}$ and 5 are equally common. In Example 8 marks of 44, 58 and 82 each occur three times. These examples show that it is possible to have more than one mode.

3   The record currently at 'Number One' or 'Top of the Pops'. The current 'Top Ten' is the list of the ten most popular records, according to sales, placed in order with the most popular first.

1.5   The median

1   171 cm   **2** Size $4\frac{1}{2}$ **3** 7 words   **4** 2 children
The items in Examples 1, 2 and 3 cannot be arranged in a 'numerical' order.

5   (a) 4   (b) 6   (c) and (d) see text

Exercise E

1   (a)   1, 2, 3, 5, 6, 7, 8. The median is 5.

(b)   1, 2, 3, 3, 4, 6. The median is 3.

(c)   1, 2, 3, 4, 4, 5, 7, 8, 9, 10. The median is $4\frac{1}{2}$.

(d)   21, 22, 24, 25, 28, 29, 30, 31. The median is $\dfrac{25+28}{2} = 26\frac{1}{2}$.

2   Only to the extent that all the data are required to obtain an order 'smallest to largest'; but the median is not affected by the actual values of the extreme numbers. In question **1**(d), the median of 1, 2, 24, 25, 28, 29, 40, 101 is still $26\frac{1}{2}$.

3   The marks, in ascending order, are as follows:

   11, 29, 29, 30, 37, 38, 44, 44, 44, 46, 48, 49, 57, 58, 58, 58, 61, 63, 70, 73, 82, 82, 82, 91

The median is $\frac{1}{2}(49+57) = 53$ marks.

1.6 The mean

1 Class XX: $5+3+3+1+4+2+2 = 20$ pupils. Class YY: 24 pupils.

2 The modes are 4 marks for XX and 7 marks for YY. These are not good representatives – e.g. no one in XX scored less than 4 marks. It would be unreasonable to say that Class YY was 'on average' 3 marks better than Class XX!

3 The medians are 6 marks (XX) and $5\frac{1}{2}$ marks (YY). These are better representatives, and it is probably reasonable to say that Class XX is about half a mark better than Class YY. But note that if just one of Class YY who had scored 5 had scored 6, then the median for YY would have been 6 also.

4 Because there are more pupils in Class YY than in Class XX.

5 For XX the answer is $6\frac{1}{2}$ or 6.5, and for YY it is $5\frac{1}{2}$ or 5.5. See the following text. (The advantage of this method is that everyone's mark counts to the same, small, extent. For the situation envisaged in question **3** above, if one pupil in YY improved from 5 to 6 marks the new mean would be $133 \div 24$, which is still approximately 5.5.)

6 No. Nine were above, and eleven below.

7 How much better, or how much worse, those pupils were than the mean.

8 18.5. The 'sum above' equals the 'sum below' because the mean is equivalent to averaging out the marks by taking from the best and giving to the worst, etc., so that eventually everyone has the same mark.

9 Yes. Sum above = sum below = 16. See Figure I.

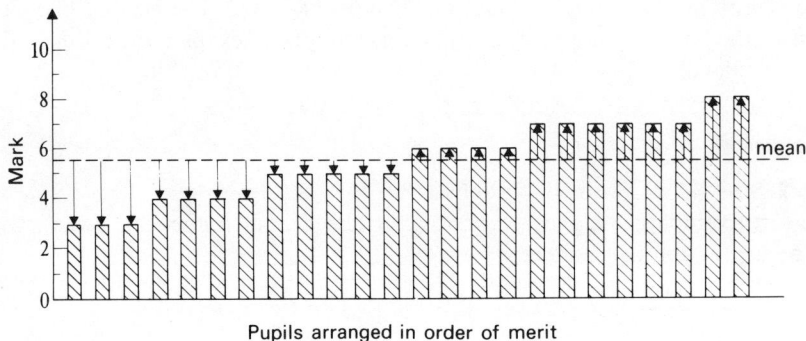

Figure I

Exercise F

1 The total number of children is $5 \times 1 + 10 \times 2 + 7 \times 3 + 2 \times 4 + 0 \times 5 + 1 \times 6 = 60$.
The number of families is $5 + 10 + 7 + 2 + 0 + 1 = 25$.
The mean number of children per family is $60 \div 25 = 2.4$.

2 $183 \div 10 = 18.3$

3 $30 \div 6 = 5$
(a) $100 + 5 = 105$ (b) $70 + 5 = 75$ (c) 1445

4 (a) $12 \times 7 = 84$ (b) 24×12 years 1 month = 290 years

5 (a) He had scored a total of $9 \times 17 = 153$ runs in the first nine innings. Hence after ten innings he had scored a total of $153 + 7 = 160$ runs for an average of $160 \div 10 = 16$ runs.
(b) For an average of 19 from eleven innings he needs a total of $11 \times 19 = 209$ runs. He must score 49 runs in his last innings.

6 (a) $9\frac{1}{4}$ words per sentence. In this example the median (7 words) is probably a 'better' average; i.e. is more typical of the sentences in the passage.
(b) The total of the marks obtained is 1284. Mean = $1284 \div 24 = 53.5$.
(c) No.

7 See Figure J.

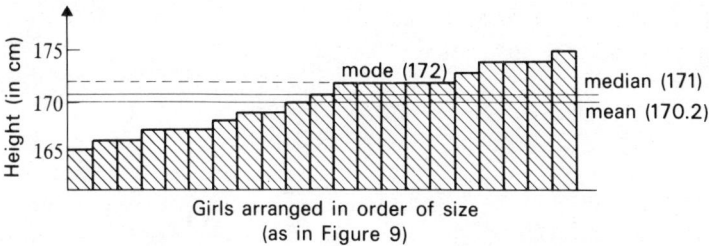

Figure J

Both the mean and the median are reasonable averages (but see the comment to questions **3** and **5** in Section 1.6).

Post-test

1 (a) See Figure K.

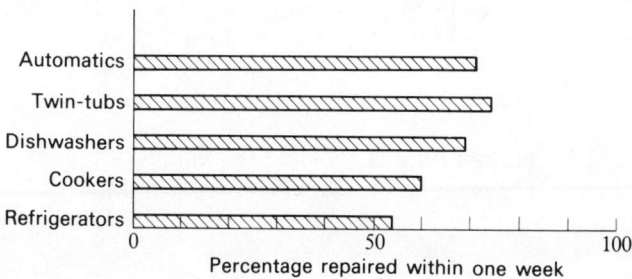

Figure K

(b) The 'sum' $(71\% + 74\% + \ldots)$ has no meaning in this example. In such cases it is not possible to draw a pie chart, which is primarily to show what fraction of the sum each item is.

2 (a)

	County Council	City Council	Water
Angle	240°	105°	15°
Percentage*	$\frac{240}{360} \times 100 = 67\%$	29%	4%
Amount*	67p	29p	4p

* to nearest whole number

23

(b) See Figure L.

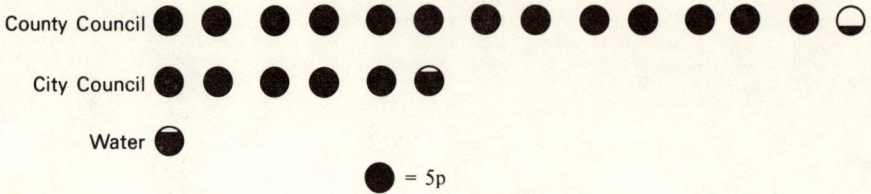

Figure L

3 (a)

Price	1p	2p	3p	4p	5p	6p	9p
Frequency	16	12	6	7	4	8	7

See Figure M.

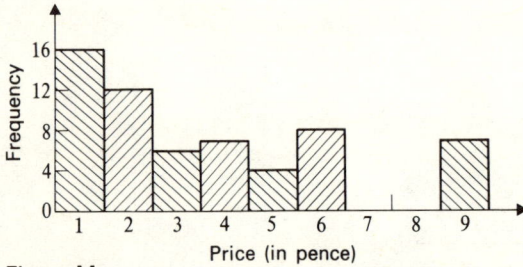

Figure M

(b) The modal price is 1p; the median price is 3p, and the mean price is 3.6p.
($16 \times 1p + 12 \times 2p + \ldots = 217p$ is the total cost. $217 \div 60 = 3.6p$.)

2 Probability

For this chapter you will need some coins, dice, coloured beads or counters and a pack of playing cards.

Objectives

This is what you should be able to do after studying this chapter.
(1) Given the appropriate data, calculate the probability based on experiment of an event occurring.
(2) List the possibility space for a trial (where appropriate), understand the meaning of the term 'equally likely outcomes', and calculate the expected probability of an event occurring.
(3) Recognise possible pitfalls in choosing samples when carrying out a survey.
(4) Use some methods of producing a random selection.

Pre-test

1 $\mathscr{E} = \{x : x < 10, x \text{ is a positive integer}\}$, and $A = \{x : x \text{ is even}\}$.
(a) List the members of A.
(b) State the value of $n(A)$.
(c) List the possible compositions of a set B, if $n(B) = 3$ and $B \subset A$.

2 Simplify. (a) $\frac{9}{15}$ (b) $\frac{42}{91}$ (c) $\frac{8}{68}$

3 In a survey on the methods of travelling to school, the tally table shown here was produced. Copy and complete the frequency table, and the frequency diagram shown in Figure 1.

Method of travel	Tally	Frequency
Walk	卌 卌 l	
Cycle	l l l l	
Bus	卌 卌 l l l	
Train	l l	
Car	卌 卌	

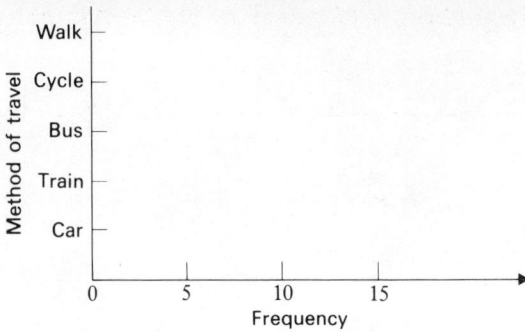

Figure 1

2.1 Experiment

'The chances of my car passing its MOT test next month are *poor*.'

'It is *almost certain* that the local football team will gain promotion at the end of the season.'

'We have a *fifty–fifty chance* of winning the toss at the beginning of the next match.'

Each of these statements concerns the probability of an event happening; and in making such comparisons we are drawing on our past experience of these events, and at the same time taking into account current circumstances that may affect the issue. (Think, for example, what considerations are made before predicting the outcome of a football match.)

In Exercise A we look at some experiments which help us to determine how probable, or how likely, certain events are under given conditions; and also what conclusions we are justified in drawing from the experiments.

Exercise A

1   Here are the results when a coin was tossed 12 times.

	Tally	Frequency
Heads	❘ ❘ ❘	3
Tails	卌 ❘ ❘ ❘ ❘	9

(a)   Does this mean that the chances of obtaining a tail are 'three times as good' as that of obtaining a head?

(b)   Approximately how many tails would you expect to obtain if the coin is tossed 1000 times?

2   Toss two coins, and record with a tally mark whether two heads, a head and a tail, or two tails, occurs. Continue this experiment until you have (a) 10 results; (b) 50 results; (c) 100 results. Illustrate your results as bar charts in each case.

(d)   From your results, what do you deduce about the relative chances of obtaining

26

a head and a tail to obtaining two tails? Which result do you think is the most reliable?

(e) What fraction of the experiments in (c) gives 2 heads?

(f) If you tossed two coins 10000 times, approximately how many times would you expect 2 tails to occur?

> **3** Put three red beads and two blue beads in a box; without looking, take one out and record its colour on a tally table. Put it back in the box, shake, and repeat. Continue until you have 60 tallies.

(a) Are you more likely to choose a red bead or a blue bead?

(b) What fraction of the time did you take out a red bead? A blue bead?

4 In games such as Monopoly, two dice are thrown together, and the score is the sum of the two numbers showing.

(a) List the possible scores that can be obtained.

(b) Is the chance of scoring 2 the same as that of scoring 8?

(c) Throw a pair of dice 100 (or preferably 200) times, and make tally and frequency tables for your results. Illustrate your results on a bar chart.

(d) Does it appear that the chances of some scores are better than others? What fractions of your results give scores of 2; 4; 7?

> **5** If, in question **3**, you had not known how many there were of each colour of bead in the box, could you have deduced it from your results?

Try the following: ask someone to put five beads (some red, some blue) in the box without telling you how many of each there are. Repeat the experiment described in question **3**, and see if you can deduce the correct answer.

2.2 Probability based on experiment

Although words and phrases such as 'most unlikely', 'probably', 'almost certainly', etc., are sufficient for some purposes, it is necessary to be more accurate on many occasions. This is particularly so when a sum of money is involved as, for example, in gambling, or in insurance. In fact, the theories behind probability were developed by Blaise Pascal in the seventeenth century to help him with his gambling at cards. Today these theories are used extensively in the fields of economics, science, industry, sociology, etc.

We shall consider how we can measure mathematically the chances of a particular event happening, based on the results of a large number of trials.

The table records the results of throwing a die (a) 48 times; and (b) 480 times.

	Score	1	2	3	4	5	6
(a)	Frequency in 48 throws	9	6	7	8	11	7
(b)	Frequency in 480 throws	83	81	79	76	80	81

We have seen in Exercise A (question **1**) that with a small number of trials no valid conclusion can be drawn. For example, a coin tossed three times might very well land head upwards each time, but we should be foolish to deduce from this that a tail would never occur. (But if in 300 throws it landed head upwards every time, we might well be justified in thinking it was a two-headed coin!) However, we find in practice that there are many occasions when, as the number of trials increases, the frequency table settles down to a 'pattern'. In the experiment described above, the proportion of threes, for example, seem to be settling down to about $\frac{1}{6}$. The actual proportion of threes in the larger experiment is expressed by the fraction $\frac{79}{480}$; and this is said to be the probability, *on the basis of this experiment*, that a three will turn up if the die is thrown again (i.e. we should expect about 79 threes in the next 480 throws of the die). This is written $p(\text{score of } 3) = \frac{79}{480}$, $p(3) = \frac{79}{480}$, or just $p = \frac{79}{480}$.

1 Using the bottom line of the table above, write down the experimental probability of throwing each of these numbers.
(a) 5   (b) 1   (c) An even number   (d) Not a 6   (e) 7

2 If a die is thrown 1000 times and a 6 is scored every time, what is the experimental probability that, if thrown again, the die will turn up a 6?
   From your result, what can you deduce about the die?

3 How often will an event occur if it has a probability of 0?
   How often will an event occur if it has a probability of 1?
   An event is the more likely to occur the nearer its probability is to what number?

The probability, based on experiment, that an event will occur after a trial has been repeated a large number of times is defined by:

$$\text{Probability (event)} = \frac{\text{The number of trials in which the event occurs}}{\text{The total number of trials}}$$

   Hence the probability of an event happening is expressed as a proper fraction, preferably in its lowest terms, or the equivalent decimal. That is, if p is the probability, $0 \leqslant p \leqslant 1$.

Exercise B

1 The political allegiances in a town are shown in the table.

Tory	35%
Socialist	45%
Liberal	20%

What is the probability that a person chosen at random is a Socialist?

2 If you choose an exercise at random from this chapter (i.e. from the pre-test, Exercises A to E, the post-test and the assignment), find out the probability of it containing the following number of questions.
(a) 5   (b) Less than 4   (c) More than 10   (d) An even number

3 A card is drawn at random from a hand of 13 cards, and its suit is noted. The card is replaced, the hand shuffled, and another card is drawn and its suit noted. This is carried out 325 times, with the results shown here.

28

Suit	Clubs	Diamonds	Hearts	Spades
Frequency	104 *2*	51 *1·5*	48 *1*	122 *2·5*
	4	*2*	*2*	*5*

(a) What is the probability of drawing a club; a diamond; a heart; a spade?

(b) How many cards of each suit does the hand appear to contain?

(c) Deal yourself a hand of 8 cards (face downwards) from a pack, and see whether you can deduce the composition of the hand in a similar way. (About 100 trials should be enough!)

4 Perform about 100 trials to find the probability of obtaining just one head when three coins are tossed together.

This is best done by tossing one coin about 300 times, and taking the results in threes; e.g. if the first few throws were *HTT THT THH TTT HTH HHT*...the beginning of the tally table would be as follows:

All heads	
Two heads and one tail	| | |
One head and two tails	| |
All tails	|

▷ 7

2.3 Expected probability

We have seen how the probability of a particular event occurring can be determined experimentally by using the results of a series of trials. However, it is very likely that someone else carrying out the same experiment would get a different set of results, and consequently a different value for the probability (although if a sufficient number of trials has been carried out the results should be approximately the same).

The probability can often be determined in a different way. Consider, for instance, the case of throwing a die. There are six possible scores, and – provided the die is 'fair' (i.e. symmetrical, and not loaded) – we expect the chance of scoring 1 to be the same as the chance of scoring 2 or 3 or 4 or 5 or 6. We say that, on the grounds of symmetry, all the six results are *equally likely*. Thus, for example, if we throw a fair die 600 times we expect the numbers 1 to 6 to occur about 100 times each. Hence the chance that any particular one number will occur is $\frac{1}{6}$.

This is called the *expected probability*.

The set of all possible outcomes of an event is often called the *possibility space* (in this example $\mathscr{E} = \{1, 2, 3, 4, 5, 6\}$); and we can define expected probability as:

$$\text{Expected probability} = \frac{\text{The number of outcomes that include the event}}{\text{The total number of possible outcomes}}$$

provided that all the possible outcomes are equally likely. (This will be considered more fully in Section 2.4.) For example, if we want the probability of scoring a 5 or a 6, the set of outcomes in which we are interested is $S = \{5, 6\}$, and the expected

probability $= \frac{n(S)}{n(\mathscr{E})} = \frac{2}{6} = \frac{1}{3}$; i.e. $p(S) = \frac{1}{3}$.

29

$\frac{13}{52} = \left(\frac{1}{4}\right)$ 0.75 0.9 Diamond $/52$.

Suppose we want to find the probability of throwing a prime number on a die.

▷ 1 What is the set of equally likely outcomes, $\mathscr{E}$, and the value of $n(\mathscr{E})$?

2 What is the set of prime scores, S, and the value of $n(S)$?

3 Hence write down the expected probability, $p(S)$.

Exercise C

▷ 1 When a card is drawn at random from a pack of 52 playing cards, find the probability of choosing one of the following.
(a) An ace (b) A heart (c) An ace or a heart (d) The ace of hearts

2 In a raffle 528 tickets are sold. If you have bought 3 of these, what is your chance of winning the first prize?

3 A penny and a fivepenny piece are tossed together.
(a) What is the probability that they will both turn up heads?
(b) What is the probability that they will not both turn up heads?
Add together the answers to (a) and (b), and explain the result.

▷ 4 A coin and a die are thrown at the same time. What is the probability of obtaining the following results?
(a) A head and a score greater than 4
(b) A tail and an even score
(c) A score of 5

5 A football match can end in three ways – a home win (H), an away win (A), or a draw (D).
(a) List the nine different ways in which two matches can end.
(b) Assuming that for any one match H, A, and D are equally likely, calculate the probability of both matches ending in a draw; and the probability of neither match ending in a draw.
(c) Are we justified in assuming that there is an equal chance of H, A, or D for a football match?

2.4 Equally likely outcomes

Two pennies are tossed simultaneously. Is it correct to give the set of equally likely possible outcomes as {2 heads, a head and a tail, 2 tails}, or not?

Look back at your results for Exercise A, question **2**. You should find that a head and a tail occurred twice as often as either two heads, or two tails. We can understand this more easily if we think of the two coins being distinguished in some way (e.g. one is a penny, and the other a fivepenny piece). We can get a head on the penny and a tail on the fivepenny; or we can get a tail on the penny and a head on the fivepenny. This shows that there are two different ways of getting a head and a tail. Listing the possible outcomes as *ordered pairs* we have:

$$\mathscr{E} = \{HH, \quad HT, \quad TH, \quad TT\}$$

Do you agree that we now have a set of equally likely outcomes?

If $S = \{HT, \quad TH\}$, then the expected probability of a head and a tail is given by $p(S) = \dfrac{n(S)}{n(\mathscr{E})} = \dfrac{2}{4} = \dfrac{1}{2}$, which should agree approximately with your experimental result in Exercise A, question 2. Look back at question 4 of that exercise to answer these questions.

1   What total scores are possible when two dice are thrown together?

2   Are they all equally likely? If not, which score do you think is the most likely?

11▷

Again, we shall make it easier if we distinguish the dice in some way (e.g. suppose one is red and the other blue) and we represent the numbers thrown by an ordered pair, the first number being the score on the red die. Thus (1, 4), (2, 3), (3, 2)...are all different outcomes although each gives a total score of 5.

3   Complete the table.

Score	2	3	4	5	...
Ordered pairs giving this score	(1, 1)	(2, 1) (1, 2)			...
Number of outcomes giving this score	1	2			...

4   How many elements does the possibility space contain?

5   State the probabilities of obtaining each of the possible scores. What is the sum of these probabilities?

12▷

Exercise D

1   Two dice are thrown. Find the probability of throwing the following.
    (a) A total of 5   (b) A total less than 5   (c) A total greater than 5

2   Instead of making a table of results for the possibility space in such experiments as throwing two dice together, we can plot the ordered pairs as coordinates. (You may, in fact, find it easier to think of the possibility space in this way.)
    (a)   Copy and complete Figure 2 to show all the possible outcomes when two dice are thrown together. For example, the cross shows the outcome (1, 4).

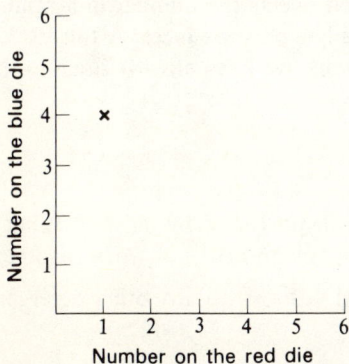

Figure 2

31

(b) Draw a loop around those crosses that represent a double such as (3, 3). Label this set as *D*. Locate similarly the set *S* = {total scores of 6}.

(c) State the probability of throwing the following: a double; a total of 6; a double or a total of 6; a double and a total of 6.

3 In a ballot to choose two representatives from Gary, Helen, Ian, John and Kate, the five names are written on pieces of paper which are put into a hat.

(a) Make a list of the ways in which two names can be drawn simultaneously from the hat.

(b) What is the probability that the two people chosen are both girls?

(c) What is the probability that they are both boys?

(d) What is the probability of one girl and one boy?

(e) What is the sum of the three probabilities in (b), (c) and (d)? Why?

4 (a) Writing *H* for heads and *T* for tails, make a list of the possible outcomes when three coins are tossed together.

(b) Calculate the probability of throwing three heads; at least two heads; at least one head; one head; no heads; four heads.

(c) In an experiment, three coins were tossed 80 times. In these trials, three heads occurred together on 8 occasions. Would you think that these coins were biased?

2.5 Random samples

1 A firm of porridge oats manufacturers wanted to find out how popular porridge was for breakfast. To do this they opened the London telephone directory at random, and contacted everyone on that page. To their delight, they found that over 90% of the people contacted had porridge for breakfast; but was this a 'random' sample?

2 If we are asked to choose a number between 5 and 12, although in theory each number has an equal probability of being selected, in practice one number is much more often chosen than any one of the others. Do you know which one?

One method of ensuring that the selection of something really is 'random' is to write each possibility on a piece of paper, and to draw one 'out of a hat' after mixing them up. Another method is to link the selection with throwing a die: for example, a correspondence can be set up between the numbers on the die and the numbers between 5 and 12. A throw of the die then selects the number in a random fashion. Obviously this has its limitations, as a die has only six faces. A third method, without these limitations, and based on work we have already done in this chapter, will be considered in question 3 below.

Exercise E

1 A small firm wants to carry out a survey on the transport used by its staff. What is wrong with the firm selecting at random three people who arrive at work together?

2 'In a random sample of 10 housewives, 9 preferred WHOOSH to any other detergent.' Is this a random selection?

3 (a) Explain the connection between the two tables shown here.

(b) How can this be used to generate a random number between 0 and 7 (inclusive)? Between 6 and 13 (inclusive)?

The ways in which three coins can land	Binary numbers
H H H	0 0 0
H H T	0 0 1
H T H	0 1 0
H T T	0 1 1
T H H	1 0 0
T H T	1 0 1
T T H	1 1 0
T T T	1 1 1

▷16

Summary

(1) The probability that something will happen can either be based on the results of a (large) number of trials, or be deduced theoretically (expected probability).

(2) If the probability is p, then $0 \leqslant p \leqslant 1$. If $p = 0$, then the event is impossible; if $p = 1$, it is a certainty. It is usual to express p either as a fraction in its lowest terms, or as the equivalent decimal.

(3) The *probability based on experiment* of an event happening is:

$$\frac{\text{The number of trials in which the event occurs}}{\text{The total number of trials that have been carried out}}$$

For example, when a coin was tossed 40 times, it came down heads on 18 occasions. The probability that the next throw will be a head, p(heads), is $\frac{18}{40}$ or $\frac{9}{20}$. By this we mean that if we carry on throwing this coin a large number of times, it should come down heads about $\frac{9}{20}$ths of the time; (but remembering that this prediction is based only on the results of the experiments that we have carried out so far).

(4) The *possibility space* of a trial (or experiment) is the set of all the equally likely possible outcomes. The decision as to whether the possibilities are equally likely is usually based on considerations of symmetry.

For example when two unbiased coins are thrown, the possibility space is the set of ordered pairs $\{HH, HT, TH, TT\}$.

(5) The *expected probability* of an event happening is:

$$\frac{\text{The number of outcomes that include the event}}{\text{The total number of possible outcomes}}$$

provided that the possible outcomes are equally likely.

For example, the probability of 2 heads when two coins are thrown $= \frac{1}{4}$.

(6) Random numbers can be generated by dice; binary numbers: drawing from a hat, or a pack of cards, and so on.

Post-test

▷ 1 How would you find the probability of a drawing-pin landing point upwards when dropped on the floor?

2 In Britain in 1966 there were 504 000 boys born, and 476 000 girls. If a Briton born in 1966 is chosen at random, what is the probability that a male is chosen?

3 A fair die is thrown. State the probability of obtaining the following.
(a) A number greater than 3   (b) A number less than 5   (c) A 7   (d) Not a 7

4 Two regular tetrahedra (symmetrical four-faced solids) are used as dice, by putting the numbers 1 to 4 on their faces. (When such a die lands, it is the hidden face that gives the score!)
(a)   Using ordered pairs, list the possible outcomes when both tetrahedra are thrown together.
(b)   What is the probability that the *product* of the two numbers is 4; 10; 12?

Assignment

1   A tyre manufacturer kept a record of the distances at which a particular brand of tyre needed replacing. The results from 800 samples are shown in the table.

Distance (km)	less than 5000	5000–9999	10 000 – 14 999	15 000 and above
Frequency	125	257	328	90

(a)   What is the probability that a tyre of this brand will need replacing before 1500 km?
(b)   If a tyre life of less than 500 km is regarded as 'unsatisfactory', state the probability that a new tyre of this brand will turn out to be unsatisfactory.

2   Would you use theory or experiment to find the probability of the following?
(a)   Tossing two coins and getting two heads.
(b)   Tossing two particular coins and getting two heads.
(c)   A person born in Great Britain in 1965 being left-handed.
(d)   Cutting a pack of cards and obtaining a Jack.

3   If the probability that a person chosen at random is left-handed is $\frac{1}{20}$, what is the probability of being right-handed?

4   A card is drawn at random from a pack of cards from which the four, five, six and seven of spades are missing. What is the probability of drawing the following?
(a) A heart   (b) A spade   (c) A six.

5   A red die and a blue die are thrown together. What is the probability of the following?
(a)   The number on the red die is greater than that on the blue.
(b)   The number on the blue die is less than 2.
(c)   The number on the red die is greater than 4 and the number on the blue die is less than 2.

34

6 If you carry out a survey amongst the people in the High Street at 11 a.m. on a Monday morning, will you have a random sample?

Answers

Pre-test

1 (a) $A = \{2, 4, 6, 8\}$ (b) $n(A) = 4$ (c) $\{2, 4, 6\}, \{2, 4, 8\}, \{2, 6, 8\}, \{4, 6, 8\}$
2 (a) $\frac{3}{5}$ (b) $\frac{6}{13}$ (c) $\frac{2}{17}$
3 See Figure A. The frequencies are 11, 4, 13, 2, 10.

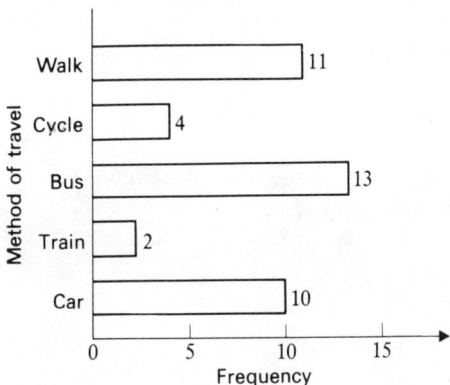

Figure A

2.1 Experiment

Exercise A

1 (a) No. The number of results (i.e. trials or experiments) is too small to make any valid prediction.
   (b) Impossible to predict, *from the evidence given*. For an unbiased coin we should expect about 500 tails.
2 If you are able to compare your results with other people (or if you can do the question several times) you will find very little similarity for (a); possibly some for (b); and quite likely some for (c). By the time you have done about 100 trials the ratio is likely to be approximately 1:2:1.
   (d) From (c) it should appear that 'a head and a tail' is about twice as likely as 'two tails'.
   (e) Approximately $\frac{1}{4}$.
   (f) 2500.
3 (a) Red (b) Your answers are likely to be about $\frac{3}{5}$ and $\frac{2}{5}$.
   Note that we have used the words 'are likely to be' here! In other words, if 100 people (say) carried out the experiment (or you did it 100 times), then 'most of the time' the results would be approximately $\frac{3}{5}$ for red and $\frac{2}{5}$ for blue; the combined results would certainly be close to $\frac{3}{5}$ and $\frac{2}{5}$ – but only about 10 people would get exactly 36 reds in 60 draws, and a few results would be 'way out'. A frequency diagram of the results would look like Figure B (overleaf).

35

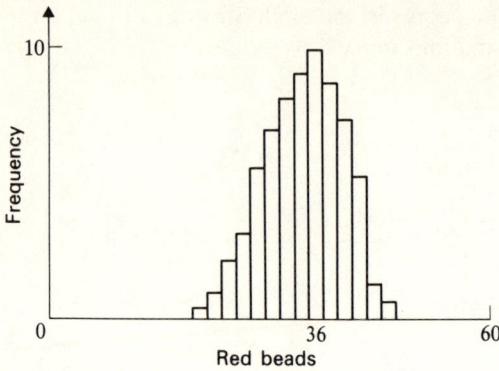

Figure B

4 (a) 2, 3, 4, ..., 12.

(b) No, 8 is more likely.

(c) Your frequency diagram will 'probably' look something like Figure C.

(d) Yes. The results should be approximately $\frac{1}{36}$; $\frac{1}{12}$; $\frac{1}{6}$.

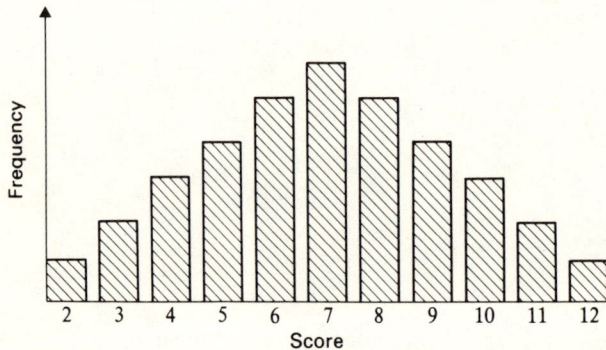

Figure C

5 Suppose your results were sixteen red beads and forty-four blue beads. These would suggest that $\frac{16}{60}$ths of the beads were red; i.e. that there were about $\frac{16}{60} \times 5 = 1\frac{1}{3}$ red beads. As the number of beads must be a whole number it is likely that there are 4 blues and 1 red; possible that there are 3 blues and 2 reds; but it is most unlikely to be any other combination.

2.2 Probability based on experiment

1 (a) $p(5) = \frac{80}{480} = \frac{1}{6}$.

(b) $p(1) = \frac{83}{480}$.

(c) $81 + 76 + 81$ ($= 238$) throws gave an even number. $p(\text{even}) = \frac{238}{480} = \frac{119}{240}$.

(d) 399 (all but 81) of the throws were not a six. Probability $= \frac{399}{480}$.

(e) No score gave a seven. $p(7) = 0$.

2 The experimental probability $= \frac{1000}{1000} = 1$.

This die is almost certainly biased! It might be very heavily weighted so that six always comes up; or it might have the number six on every face!

36

3 $p = 0$ means that the event *never* occurs; i.e. it is impossible.

$p = 1$ means that the event *always* occurs; i.e. it is a certainty.

The nearer a probability is to 1, the more likely is the event to occur.

Exercise B

1 Per cent means 'in every hundred', therefore there are 45 Socialists in every hundred citizens (on average).

$p(\text{Socialist}) = \frac{45}{100} = \frac{9}{20}$ (or 0.45).

2 The number of questions in the exercises are 3, 5, 4, 5, 4, 3, 4 and 6.

(a) Two of the eight exercises have five questions: $p = \frac{2}{8} = \frac{1}{4}$.

(b) $\frac{2}{8} = \frac{1}{4}$ (c) 0 (d) $\frac{4}{8} = \frac{1}{2}$

3 (a) $p(\text{club}) = \frac{104}{325} = \frac{8}{25}$; $p(\text{diamond}) = \frac{51}{325}$; $p(\text{heart}) = \frac{48}{325}$; $p(\text{spade}) = \frac{122}{325}$.

(b) This is similar to question **5** of Exercise A. 4 clubs, 2 diamonds, 2 hearts, 5 spades. (For example, the number of spades is about $\frac{122}{325} \times 13 \approx 5$.)

4 Your answer should be approximately $\frac{3}{8}$.

2.3   Expected probability

1 $\mathscr{E} = \{1, 2, 3, 4, 5, 6\}$; $n(\mathscr{E}) = 6$.

2 $S = \{2, 3, 5\}$; $n(S) = 3$.

3 $p(S) = \frac{3}{6} = \frac{1}{2}$.

Exercise C

1 (a) $\frac{4}{52} = \frac{1}{13}$.

(b) $\frac{13}{52} = \frac{1}{4}$.

(c) There are 4 aces, and 12 *other* hearts in the pack. $n(S) = 16$; $p(S) = \frac{16}{52} = \frac{4}{13}$.

(d) There is only one 'ace of hearts' in the pack. $p = \frac{1}{52}$.

2 $\frac{3}{528} = \frac{1}{176}$.

3 There are four possible outcomes: ($1p = H$, $5p = H$), ($1p = H$, $5p = T$), ($1p = T$, $5p = H$), ($1p = T$, $5p = T$). It seems reasonable to suppose that these are equally likely.

(a) $\frac{1}{4}$ (b) $\frac{3}{4}$.

The sum is 1, because it is a certainty that one or other of (a) and (b) will happen.

4 There are now 12 possible outcomes, as shown diagrammatically in Figure D, and these are all equally likely on the grounds of symmetry.

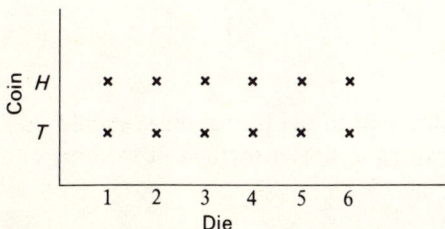

Figure D

(a) $S = \{H5, H6\}$; $n(S) = 2$; $p(S) = \frac{2}{12} = \frac{1}{6}$.

(b) $S = \{T2, T4, T6\}$; $n(S) = 3$; $p(S) = \frac{3}{12} = \frac{1}{4}$.

(c) $S = \{H5, T5\}$; $n(S) = 2$; $p(S) = \frac{2}{12} = \frac{1}{6}$.

5 (a)

| First match | H H H | A A A | D D D |
| Second match | H A D | H A D | H A D |

(b) $\frac{1}{9}$; $\frac{4}{9}$.

(c) No; in practice, a home win is more likely.

2.4 Equally likely outcomes

1 2, 3, 4, ..., 12.

2 No. Seven is the most likely score.

3 See the completed table.

Score	2	3	4	5	6	7	8	9	10	11	12
Ordered pairs giving this score	(1, 1)	(2, 1) (1, 2)	(3, 1) (2, 2) (1, 3)	(4, 1) (3, 2) (2, 3) (1, 4)	(5, 1) (4, 2) (3, 3) (2, 4) (1, 5)	(6, 1) (5, 2) (4, 3) (3, 4) (2, 5) (1, 6)	(6, 2) (5, 3) (4, 4) (3, 5) (2, 6)	(6, 3) (5, 4) (4, 5) (3, 6)	(6, 4) (5, 5) (4, 6)	(6, 5) (5, 6)	(6, 6)
Number of outcomes giving this score	1	2	3	4	5	6	5	4	3	2	1

4 36 elements.

5 Probabilities: $p(2) = \frac{1}{36}$, $p(3) = \frac{2}{36} = \frac{1}{18}$, $p(4) = \frac{3}{36} = \frac{1}{12}$, $p(5) = \frac{4}{36} = \frac{1}{9}$, $p(6) = \frac{5}{36}$, $p(7) = \frac{6}{36} = \frac{1}{6}$, $p(8) = \frac{5}{36}$, $p(9) = \frac{4}{36} = \frac{1}{9}$, $p(10) = \frac{3}{36} = \frac{1}{12}$, $p(11) = \frac{2}{36} = \frac{1}{18}$, $p(12) = \frac{1}{36}$. The sum of the probabilities = 1.

Exercise D

1 The following results come from the answer to question **3** above.

(a) $\frac{4}{36} = \frac{1}{9}$

(b) $\dfrac{1+2+3}{36} = \dfrac{6}{36} = \dfrac{1}{6}$

(c) $\dfrac{5+6+5+4+3+2+1}{36} = \dfrac{26}{36} = \dfrac{13}{18}$

As one of these three events must occur, the sum of the probabilities should be 1. This can be used either as a check on (c), or as a shorter method of working out (c) (i.e. the answer to (c) is $1 - (\frac{4}{36} + \frac{6}{36}) = \frac{26}{36} = \frac{13}{18}$).

2 (a) and (b) see Figure E.

(c) $n(\mathcal{E}) = 36$; $n(D) = 6$; $n(S) = 5$; $n(D \cup S) = 10$; and $n(D \cap S) = 1$.

Hence $p(D) = \frac{6}{36} = \frac{1}{6}$; $p(S) = \frac{5}{36}$; $p(D \cup S) = \frac{10}{36} = \frac{5}{18}$; $p(D \cap S) = \frac{1}{36}$.

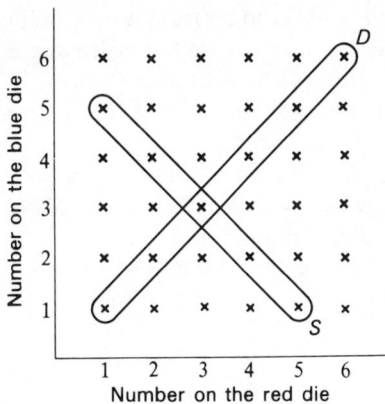

Figure E

◄4► 3   (a)   Using initial letters, we have *GH, GI, GJ, GK, HI, HJ, HK, IJ, IK, JK* (total 10). (Note that the names are drawn simultaneously, so these pairs are not ordered.)

      (b)   *H* and *K* are the girls: *HK* is the only pair which is both girls, hence $p(\text{both girls}) = \frac{1}{10}$.

      (c)   Three pairs – *GI, GJ, IJ* – are both boys, so $p(\text{both boys}) = \frac{3}{10}$.

      (d)   The mixed pairs are *GH, GK, IH, IK, JH* and *JK*, so $p(\text{mixed pair}) = \frac{6}{10}$ or $\frac{3}{5}$.

      (e)   The sum of these probabilities is 1, since a choice of two people must be one of these combinations.

  4   Compare question **4** of Exercise B.

      (a)   *HHH, HHT, HTH, HTT, THH, THT, TTH, TTT* (total 8).

      (b)   As the above outcomes are equally likely, the probabilities are as follows: $p(\text{three heads}) = \frac{1}{8}$; $p(\text{two heads at least}) = \frac{1}{8} + \frac{3}{8} = \frac{1}{2}$; $p(\text{at least one head}) = \frac{1}{8} + \frac{3}{8} + \frac{3}{8} = \frac{7}{8}$ (i.e. one head, or two heads, or three heads); $p(\text{one head}) = \frac{3}{8}$; $p(\text{no heads}) = \frac{1}{8}$; $p(\text{four heads}) = 0$.

      (c)   The probability, therefore, of three heads is $\frac{8}{80}$ or $\frac{1}{10}$. The expected probability for unbiased coins is $\frac{1}{8}$; and so this is a reasonable result to expect from unbiased coins.

2.5   Random samples

◄5► 1   The firm probably opened the telephone directory at 'Mac–' or 'Mc–'! Random samples should be scattered. In a directory it could be the first name on each page – but remember that a telephone directory in itself is not a random sample of the population as it does not include those who cannot afford, or do not want, a telephone.

  2   In practice, 7 is the number chosen most frequently. Superstition?

Exercise E

◄6► 1   They probably all used the same method of transport, particularly if they came by bus or train.

39

2 Very unlikely. A sample of 10 is too small, and is likely to be from a very limited cross-section of the public. They may well have been interviewed after leaving a shop which was running a publicity campaign to sell 'Whoosh' at cut prices.

3 (a) There is a one-to-one correspondence between the tables; i.e. if H is replaced by 0 and T by 1 then the first table becomes the second.

    (b) Toss a coin three times, and translate the outcomes into binary (as in (a) above) and then denary. For example, suppose the coins came down HTH. This is 010 in binary, or 2 in denary, and our random number is 2.

        For numbers between 6 and 13 adopt the same procedure, but add on 6 at the end to give the range 6 to 13. For example, THT is 101 in binary, which is 5 in denary. Adding on 6, the random number is 11.

Post-test

1 Drop a single drawing-pin on the floor a large (100, say) number of times, and record how many times it lands points upwards. The experimental probability will be:

$$\frac{\text{Number of times it lands point upwards}}{\text{Total number of trials carried out}}$$

(Dropping a box of pins would be quicker, but less reliable because of the interaction of the pins in the middle of the 'bunch'.)

2 $\dfrac{504\,000}{504\,000 + 476\,000} = \dfrac{504}{980} = \dfrac{18}{35} (= 0.514\ldots)$

3 (a) $\frac{3}{6}$ or $\frac{1}{2}$ (b) $\frac{4}{6}$ or $\frac{2}{3}$ (c) 0 (d) 1

4 (a) (1, 1), (1, 2), (1, 3), (1, 4), (2, 1), (2, 2), ... (total 16).

    (b) A product of 4 is given by (1, 4), (2, 2) and (4, 1), so $p = \frac{3}{16}$; a product of 10 is not given by any of the pairs, so $p = 0$; a product of 12 is given by (3, 4) and (4, 3), so $p = \frac{2}{16} = \frac{1}{8}$.

3 Further Statistics

Objectives

This is what you should be able to do after studying this chapter.
(1) Group a large collection of data into *classes*, and know the meaning of *class interval*, *class mark*, and *grouped frequency table*.
(2) Represent a grouped frequency table by a bar chart.
(3) Find the mean of data represented by a grouped frequency table.
(4) Understand what is meant by *cumulative frequency*, and be able to (a) make cumulative frequency tables, (b) draw cumulative frequency curves, and (c) use these to estimate the median.
(5) Understand what is meant by *quartiles* and *inter-quartile range*, and estimate their values from the cumulative frequency curve.

Pre-test

> 1 Find the mean and median of these sets of readings.
(a) 1, 6, 7, 9, 12
(b) ⁻5, 8, 11, 0, ⁻9

2 Find the mean, mode and median for the following frequency table (x is the measurement being recorded, e.g. heights of seedlings, and f is the frequency, i.e. the number of times that each measurement occurs).

x	54	57	59	61	68
f	4	7	8	5	6

3 Represent the following data as a frequency diagram (bar chart), and find the mode and the median. The data shows the heights (to the nearest 3 cm) of a sample of 124 fir tree seedlings.

Height	33	36	39	42	45	48	51	54	57	60	63
Number	5	9	14	18	20	17	18	13	6	3	1

3.1 Grouped frequency tables

Working through the pre-test will have given you more practice in representing data by a bar chart, and in calculating various averages. But what do we do when the data become more numerous?

For example: sixty 15-year-old boys were tested to find their resting pulse rates. The results (in beats per minute) were as follows:

```
72 70 66 74 81   70 74 53 57 62   78 67 75 80 84
58 92 74 67 62   91 73 68 65 80   76 74 65 84 79
61 72 72 69 70   82 79 71 86 77   69 72 56 70 62
80 76 72 68 65   76 56 86 63 73   70 75 73 89 64
```

▷ 1 The readings range from 53 to 92, so there cannot be many readings having the same value. Make up the frequency table to confirm this.

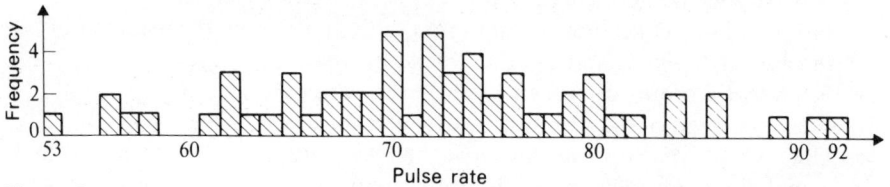

Figure 1

The frequency diagram corresponding to this frequency table is not very helpful in giving a picture of the data: it is 'messy' and has gaps in it (Figure 1).

So what we do is to form a *grouped frequency table* by grouping the individual pulse rates into *classes* such as 50–54, 55–59, 60–64, etc. We then consider the number of boys in each class. (The *class interval* in this case is 5; had we taken classes of 50–59, 60–69, etc., the class interval would have been 10.)

▷ 2 From the frequency table that you have just made, complete this grouped frequency table.

Class	50–54	55–59	60–64	65–69	70–74	75–79	80–84	85–89	90–94
Frequency	1	4							

3 Now copy and complete the bar chart in Figure 2. Note that, if you are being accurate, the bar for the class 50–54 starts at $49\frac{1}{2}$ and finishes at $54\frac{1}{2}$.

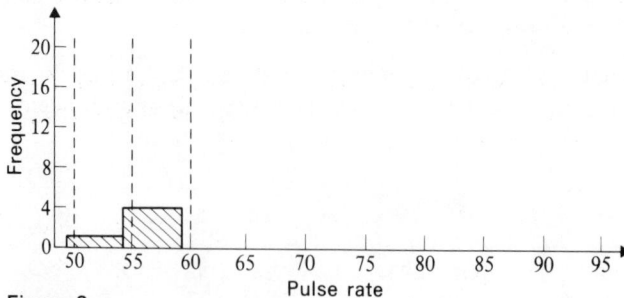

Figure 2

42

1 Repeat the table and the bar chart for the pulse rate data using class intervals of 10, starting with 50–59.

2 In a small factory employing part-time workers the weekly wages, in pounds, of the workers were as follows:

22, 24, 29, 29, 30, 32, 32, 32, 34, 35, 35, 35, 35, 40, 40, 40, 40, 40, 45, 45, 45, 50, 55, 60

(a) Compile three frequency tables for these wages, first in classes of 5, starting with 21–25; then in classes of 10, starting with 21–30; then in classes of 20, starting with 21–40.

(b) Draw the bar charts corresponding to these three frequency tables. Which chart shows the *spread* of the wages most clearly?

3.2 The mean of a grouped frequency table

The bar chart in Figure 3 represents the results of an examination. It is not easy to find the mean or median from such a bar chart. For example, although we can see that 11 pupils earned marks in the 40–59 class, we do not know their individual marks.

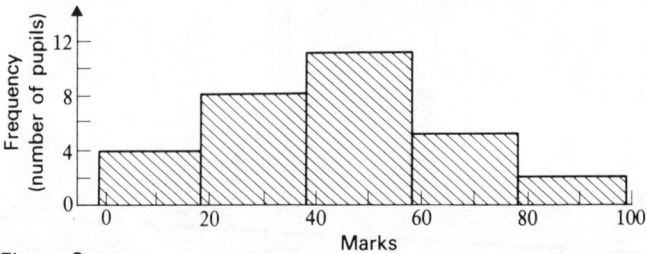

Figure 3

We get round this difficulty by assuming that these eleven pupils were 'evenly spread out' in this class; i.e. that their average mark was 50. (Strictly this should be $49\frac{1}{2}$, but because of our assumptions and the fact that our answer is going to be an estimate and therefore only approximate, it is good enough to take it as 50. If you think the accuracy justifies it, it is mathematically sound to use 50 in the working, and then subtract $\frac{1}{2}$ at the end.) With 50 as the average mark, the sum of the marks of these pupils was $11 \times 50 = 550$. This 'halfway mark' is often called the *class mark*.

1 Complete this table for the data in Figure 3. Find the sums of the last two columns and hence make an estimate of the mean.

Class interval	Class mark (*m*)	Frequency (*f*)	Estimated sum of marks in the class ($m \times f$)
0–19			
20–39			
40–59	50	11	$11 \times 50 = 550$
60–79			
80–99			

As we are now making some assumptions, and therefore approximations, it is not sensible to be as accurate in the answer as previously. When using grouped data the answer should be given to the same accuracy as the data; e.g. in this example the estimate of the mean should be given to the nearest whole number. (Question **1** of Exercise B will give you an idea of the accuracy.)

Exercise B

▷ **1** Calculate the mean for the pulse rate data in Section 3.1 in the following ways.
(a) From the individual results.
(b) From a class interval of 5 (use your results from Section 3.1, question **2**).
(c) From a class interval of 10 (use your results from Exercise A, question **1**).
Which of the last two is the better estimate for (a)?

▷ **2** Figure 4 shows the amount of money collected by 60 children on a charity walk. Estimate the mean amount collected per child.

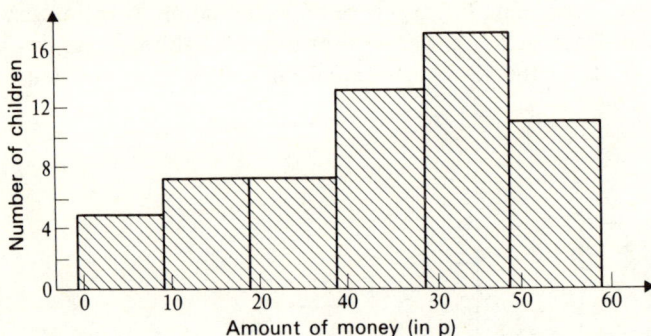

Figure 4

3 The number of words in 100 sentences of a book were counted and grouped, giving this table. Estimate the mean number of words per sentence.

Number of words	1–5	6–10	11–15	16–20	21–25	26–30	31–35
Frequency	15	27	32	15	7	3	1

3.3 Mode, median, and cumulative frequency

The modal class

▷ **1** With a large quantity of data it is sometimes misleading to quote a single measurement as the mode. Look back at Figure 1: what is the modal pulse rate?

2 However, there is some value in stating a *modal class*. Look at the two frequency diagrams that you have drawn for the grouped data for the pulse rates (Figures B and C). State the modal class for each. Which is the more informative?

44

The median

It is not so easy to find the median for grouped data. Look back at Figure 3: how can we work out a median for this example?

There are 30 pupils altogether, and so the median is a mark in between the marks of the 15th and 16th pupils. (As represented on a frequency diagram, they are in order of merit already.) The first twelve pupils scored less than 40 marks, so the 15th, 16th and median marks are in the 40–59 class. As when we worked out the mean, we assume that in any interval the marks are spread out evenly throughout the class. Figure 5 shows how we can estimate the median on this assumption.

$$\text{Median} = 39\tfrac{1}{2} + \tfrac{3}{11} \text{ of } 20 \approx 39.5 + 5.5 = 45$$

Accuracy to the nearest whole number is as far as it is sensible to go. In fact, if the data is in class intervals of 5 or less, it is probably better to quote a *median class* instead of a single value.

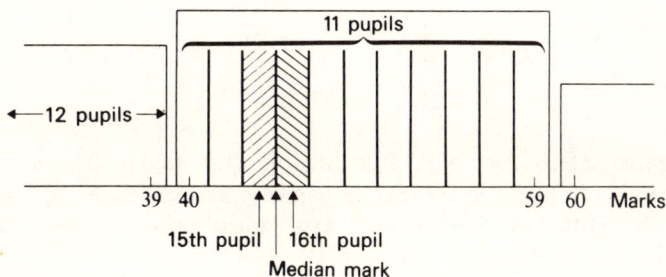

Figure 5

The cumulative frequency curve

The calculation of the median can sometimes be awkward; an alternative method is to construct the *cumulative frequency table* and draw the *cumulative frequency curve*. A cumulative frequency is a running total; for example, the 23 in the table shows that a *total* of 23 pupils scored less than 60 marks. It is the sum of the frequencies so far.

> **3** Complete the table using the data in Figure 3.

Class	Frequency	Cumulative frequency
0–19	4	4
20–39	8	$4+8 = 12$
40–59	11	$12+11 = 23$
60–79		
80–99		

We can now answer these questions.

4 How many pupils scored 39 marks or less (i.e. less than 40)?

5 How many scored 60 or more?

45

However, it is no easier yet to say how many scored 50 or more, or to state the median mark. But if we plot the cumulative frequencies on a graph it does become easier (Figure 6).

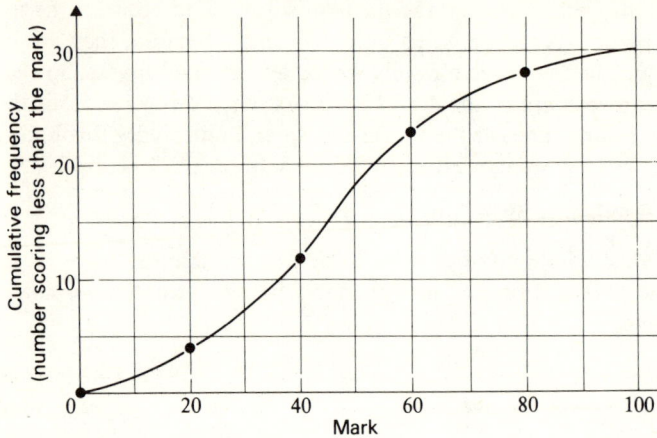

Figure 6

Note that the cumulative frequency of 23 (for example) is plotted at the *end* of its class interval, at either $59\frac{1}{2}$ or 60, to show that 23 pupils scored less than $59\frac{1}{2}$ (or 60) marks. (It could also be plotted at 59 – and the *y*-axis named accordingly – to show that 23 pupils scored *59 marks or less*.)

How many pupils scored 55 or more marks? This is equivalent to asking 'How many pupils scored less than 55?' To answer this question; follow these steps, as shown in Figure 7.

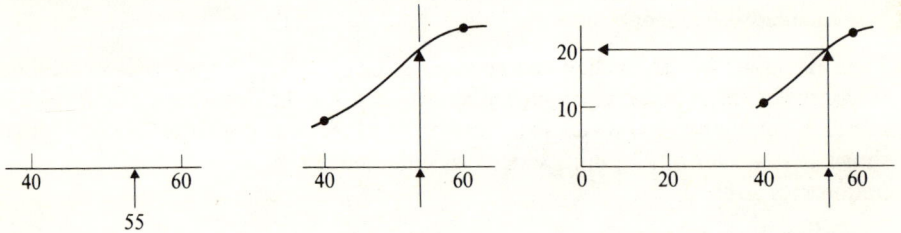

Figure 7

Start at 55 on the 'marks' axis (left-hand diagram); go up the page to meet the curve (centre); go across the page to meet the 'c.f.' axis (right-hand diagram). From this we see that *20 pupils* scored less than 55 marks; hence 10 pupils (30 − 20) must have scored 55 marks or more.

What is the median mark? This is equivalent to asking 'What is the smallest mark that was such that 15 pupils scored less than it?' Follow these steps as shown in Figure 8: start at 15 on the 'c.f.' axis; go across to the curve; go down to the 'marks' axis; read off the mark.

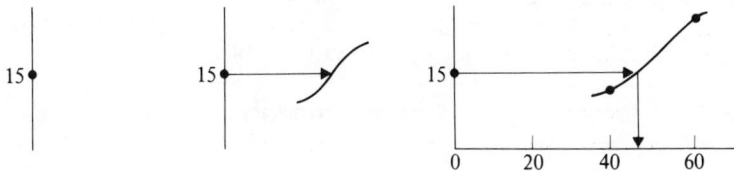

Figure 8

Thus the median is about 45 marks. Unless you have drawn a very large and accurate cumulative frequency curve, this method may seem 'less accurate' than the calculation method – but remember that it is only an estimate in either case, and so cannot be very accurate!

Non-integral data

So far in our statistical work, the data that we have used has been collected from a background of possibilities which includes only whole numbers (e.g. marks, number of children, number of words). When the information is to be grouped, it is easy to decide into which group each statistic is to be placed, and the boundaries between groups are not critical.

However, if the data we collect comes from a 'continuous variable' (i.e. every point on the number line is a possible measurement, as in the length of a piece of string, a person's height, the mass of a chicken, etc.) we have to be a little more careful.

If, for example, we have the masses of 20 battery hens we have first to decide the degree of accuracy to which we are going to work. Suppose we choose to measure to the nearest 100 grams, the results (in kg) might be as follows:

$$2.7 \quad 2.6 \quad 2.7 \quad 2.9 \quad 3.1 \quad 2.4 \quad 2.3 \quad 2.7 \quad 2.8 \quad 3.7$$
$$3.1 \quad 2.7 \quad 2.5 \quad 2.6 \quad 2.5 \quad 2.9 \quad 3.0 \quad 2.5 \quad 2.6 \quad 2.9$$

These results could be grouped as in the table.

Mass (in kg)	Frequency	Cumulative frequency
2.0–2.4	2	2
2.5–2.9	14	16
3.0–3.4	3	19
3.5–3.9	1	20

A slight difficulty arises, however, when we consider plotting the points on the cumulative frequency diagram. Where do we plot a cumulative frequency of 16, say? Since a mass of 2.947 kg would be included in the 2.5–2.9 class (as 2.947 is 2.9 correct to 1 decimal place), but a mass of 2.952 kg would be included in the 3.0–3.4 class, it should be clear that the cumulative figure for 2.5–2.9 (i.e. 16) includes all measurements up to, but not including, 2.9500... Hence we should plot 16 against a mass of 2.95 kg precisely: this signifies that 16 (and no more) hens had a mass of less than 2.95 kg, and all the others had a mass of 2.95 kg or more.

47

Exercise C

▷ 1   (a)   Complete the cumulative frequency column in the table.

Marks in a test	Frequency	Cumulative frequency
1–10	3	
11–20	17	
21–30	41	
31–40	85	
41–50	97	
51–60	115	
61–70	101	
71–80	64	
81–90	21	
91–100	6	

(b)   Draw the cumulative frequency diagram.

(c)   Use the diagram to find the median mark; the number of candidates who scored 35 or less; the number of candidates who scored 75 or more; the number of candidates who passed, if the pass mark was 45.

(d)   Calculate 60% of the total number of candidates, and find what the pass mark would have to be for just 60% of the candidates to pass.

▷ 2   A competition at a fete involved guessing the number of marbles in a glass jar. 499 people took part, and the table shows how their guesses were grouped.

Number of marbles	Number of competitors
251–300	7
301–350	28
351–400	83
401–450	170
451–500	99
501–550	72
551–600	30
601–650	10

(a)   Draw the cumulative frequency diagram.

(b)   If the correct answer was 447, estimate how near the median mark was to it. Did competitors tend to 'over-guess' or 'under-guess'?

3

Mass (in grams, correct to the nearest mg)	Frequency
11.335–11.339	7
11.340–11.344	29
11.345–11.349	43
11.350–11.354	16
11.355–11.359	5

The table shows the results when one hundred 10p coins were weighed. Draw the cumulative frequency diagram for this data, and read off the median mass of a 10p coin.

11▷

3.4 Quartiles and inter-quartile range

Quartiles

Look at your answers to question **1** of Exercise C. A sketch of the cumulative frequency diagram is shown in Figure 9. With the help of this diagram it is fairly easy to estimate the median. Two other values which are often useful are the quarter-way and three-quarter-way marks. In Figure 9, these would be the marks of the $137\frac{1}{2}$th ($\frac{1}{4} \times 550$) or 138th, and $412\frac{1}{2}$th ($\frac{3}{4} \times 550$) or 413th, persons. These marks are called *quartiles*, as they divide the data (in this case 550 scores) into four equal parts. To distinguish between them they are called the *lower quartile* and *upper quartile*, as shown in Figure 9.

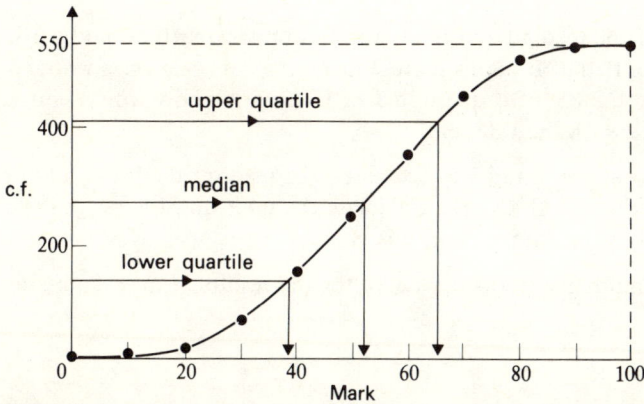

Figure 9

> **1** From your answer to question **1** of Exercise C (Figure F) read off the lower and upper quartiles.

2 What is another name for the *middle quartile*?

3 Subtract the lower quartile from the upper quartile.

12▷

Inter-quartile range

The answer to question **3** above is called the *inter-quartile range*; it is a measure applied to the middle half of the population, and it gives an idea of how spread out the data is – but it ignores the extremes (the top and bottom 25%).
 On the grouped frequency bar chart in Figure 10 (over) the quartiles and median are marked. Note that by virtue of their definitions they divide the bar chart into four parts of equal *area*. The precise position (and hence value) of the quartiles can be worked out in the same way as for the median; but again it is much easier, and usually quite accurate enough, to read them from the cumulative frequency diagram.

49

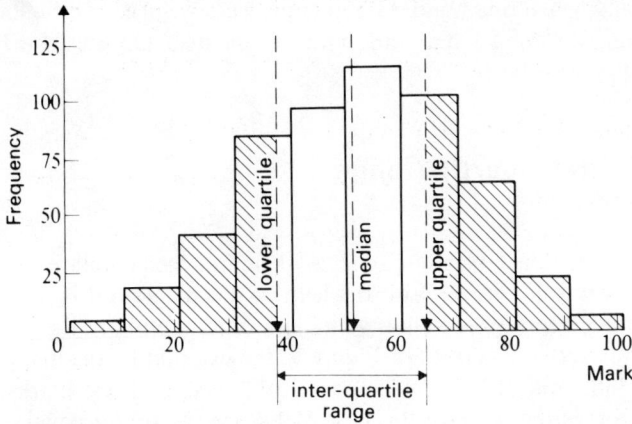

Figure 10

Exercise D

▷ 1   From Figure 6 read off the median and the lower and upper quartiles, and work out the inter-quartile range. If the 30 pupils represented here were one class whose results were included in the data for question **1** of Exercise C, how would you say that this class compared with the total entry?

2   Use your answers to questions **2** and **3** of Exercise C to read off the lower and upper quartiles in each case (Figures G and H). Calculate the inter-quartile range for each example, and say briefly what information this gives.   ▷13▷

▷ 3   Figure 11 shows the cumulative frequency curve for the results of an examination.

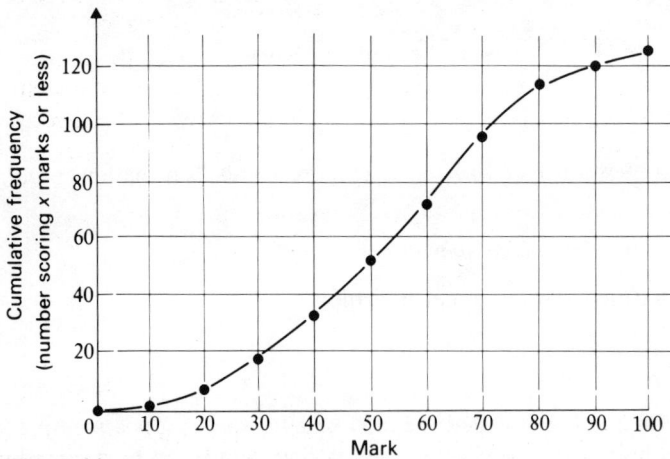

Figure 11

Use the graph to answer these questions.

(a)   What was the total number of candidates?
(b)   Find the median mark and the inter-quartile range.
(c)   How many passed if the pass mark was 40?

50

4 The table shows the results of a check on the speeds (measured to the nearest 10 k.p.h.) of 200 vehicles.
  (a) Draw the cumulative frequency curve and estimate the median and inter-quartile range.
  (b) Estimate the percentage of vehicles travelling at over 100 k.p.h.

Speed	40	50	60	70	80	90	100	110	120
Frequency	1	2	11	14	38	47	51	32	4

14▷

Summary

(1) When we have a lot of data or information to represent on a frequency diagram (bar chart) this data is placed in groups or *classes*. Usually a *class interval* is chosen so that the number of classes is between 5 and 10. (In the example in Section 3.1 the information is placed in 9 classes, using a class interval of 5.)

(2) The mean of data presented in grouped frequencies is estimated by using the assumption that the middle value of the class interval (the *class mark*) is representative of all the data in that interval (Section 3.2).

(3) The *cumulative frequency curve* is the graph of the running totals of the frequencies. It enables us to read the median (middle value) and quartiles easily.

(4) The *quartiles* are the data values which are in the one-quarter and three-quarter positions when the data is arranged in order (i.e. they divide the data into four parts with an equal number of measurements in each).
    For example, if the population (total number of items) is 100, the lower quartile is between the 25th and 26th items, the median is between the 50th and 51st items, and the upper quartile is between the 75th and 76th items.

(5) The *inter-quartile range* is the difference between the values of the upper and lower quartiles. It gives an indication of the spread of the data.

(6) The *modal class* is the class with the largest frequency. (It is possible to consider the mode as being the middle value of this class.)

Post-test

▷ **1** (a) Group the following data into classes of 5, starting with 0–4.

            26 17 42 28 23   17 39 3 38 10   44 25 4 10 33
            11 38 23 42 29   40 37 18 6 22   18 15 32 32 15
            12 37 38 37 12   29 12 18 27 30   30 12 40 45 4
            26 24 22 33 17   8 3 7 42 39   6 41 20

  (b) Draw the bar chart for the grouped frequencies.
  (c) Find the mean of the data, first by considering individual results; and then by using the grouped frequencies.

2   The masses of 100 parcels are grouped to give the *cumulative* frequencies shown in the table. Draw the cumulative frequency diagram, and use it to estimate the median and the inter-quartile range.

Mass	Cumulative frequency
Less than 2 kg	10
Less than 3 kg	27
Less than 4 kg	53
Less than 5 kg	74
Less than 6 kg	93
Less than 7 kg	100

3   The number of league appearances for the 24 members of a football club during a season were given as follows:
7, 3, 28, 5, 16, 45, 18, 1, 44, 2, 27, 6, 43, 45, 3, 34, 1, 39, 4, 43, 45, 2, 44, 6
(a)   Calculate the mean number of appearances.
(b)   Compile a grouped frequency table that has 5 classes.
(c)   Draw a bar chart to show these grouped frequencies, and estimate the mean of the grouped frequencies.
(d)   State the median number of appearances.

15

Assignment

1   A charity collects used stamps on pieces of envelope, and sells them in 100 gram packets. Here is the number of stamps in 30 packets.

219 250 231 277 204   233 271 245 211 293
257 299 198 244 271   266 240 222 301 221
250 249 269 257 201   209 236 333 247 239

(a)   Compile a grouped frequency table, using the seven classes: under 200, 200–219, 220–239, 240–259, 260–279, 280–299, 300 and over.
(b)   Estimate the mean and median number of stamps per packet. Which would you quote as 'average number' of stamps per packet?

2   A $2\frac{1}{2}$ kg sack of new potatoes is found to contain 50 potatoes whose mass distribution (correct to the nearest gram) is shown here.

Mass	1–30	31–60	61–90	91–120	121–150
Frequency	14	23	8	4	1

(a)   Estimate the mean mass of a potato.
(b)   Draw the cumulative frequency diagram.
(c)   State the median class, and estimate the median mass.

3 The table shows the length of life (in hours) of 400 electric light bulbs.

Life (hours)	Number of bulbs	Life (hours)	Number of bulbs
200–299	10	700–799	76
300–399	26	800–899	62
400–499	32	900–999	34
500–599	60	1000 or over	12
600–699	88		

(a) Draw the cumulative frequency curve and estimate the median life of a bulb.

(b) How many bulbs lasted less than 450 hours?
How many lasted longer than 725 hours?

(c) What is the value of x if the statement 'minimum life x hours' is to be true 95% of the time?

4 The table gives the information about the diameters (in mm) of the pearls in three necklaces of 100 pearls each. On the basis of this information, state which necklace you would choose to satisfy these conditions.

(a) The necklace with the pearls most exactly matched for size.

(b) The longest necklace.

(c) The necklace with at least 50 pearls of more than 7.4 mm diameter.

Necklace	Mean diameter	Median diameter	Inter-quartile range
A	7.1	7.3	1.6
B	7.4	7.5	1.4
C	7.5	7.3	1.7

Answers

Pre-test

1 (a) Mean $= (1+6+7+9+12) \div 5 = \frac{35}{5} = 7$.
Arranged in order the numbers are 1, 6, 7, 9, 12.
The median (middle number) is 7.

(b) Mean $= (19-14+0) \div 5 = 1$. Median $= 0$ ($-9, -5, 0, 8, 11$).

2 The mode (the commonest value of x) is 59.
The median (the middle value when arranged in order) is 59. (There are 30 values altogether. The 15th and 16th are both 59, hence the median is 59.)

The mean $= \dfrac{4 \times 54 + 7 \times 57 + \ldots + 6 \times 68}{30} = \dfrac{1800}{30} = 60$.

3 See Figure A. Note that as the measurements are to the nearest 3 cm, the '33' bar includes heights between $31\frac{1}{2}$ and $34\frac{1}{2}$ cm. (An alternative way of scaling the axis across the page is shown in the diagram.)

53

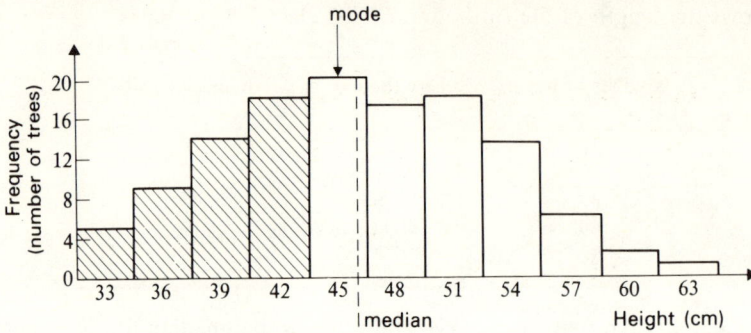

Figure A

The mode is 45 cm (or the 44–46 group).

The median is a height between the 62nd and 63rd tree (when arranged in order). The shaded bars account for 46 trees; of the 20 trees in the '45' group it is reasonable to suppose there are about 6 of 44 cm, 7 of 45 cm and 7 of 46 cm; and so the 62nd, 63rd, and therefore the median, would be 46 cm.

3.1 Grouped frequency tables

1

Pulse rate	53	56	57	58	61	62	63	64	65	66	67	68	69	70	71
Frequency	1	2	1	1	1	3	1	1	3	1	2	2	2	5	1

Pulse rate	72	73	74	75	76	77	78	79	80	81	82	84	86	89	91	92
Frequency	5	3	4	2	3	1	1	2	3	1	1	2	2	1	1	1

2

Class	50–54	55–59	60–64	65–69	70–74	75–79	80–84	85–89	90–94
Frequency	1	4	6	10	18	9	7	3	2

3   See Figure B.

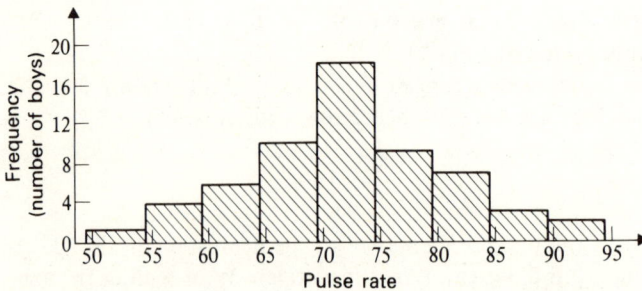

Figure B

Exercise A

Class	50–59	60–69	70–79	80–89	90–99
Frequency	5	16	27	10	2

See Figure C.

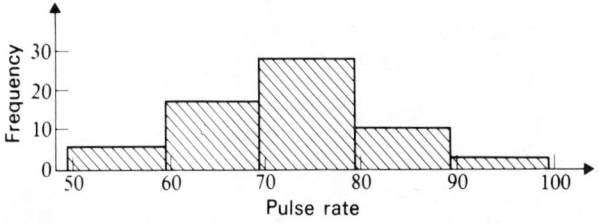

Figure C

2 (a)

Wage (£)	21–25	26–30	31–35	36–40	41–45	46–50	51–55	56–60
Frequency	2	3	8	5	3	1	1	1

Wage (£)	21–30	31–40	41–50	51–60
Frequency	5	13	4	2

Wage (£)	21–40	41–60
Frequency	18	6

(b) See Figure D. Note that the scales are chosen so that the total area under each graph is the same.

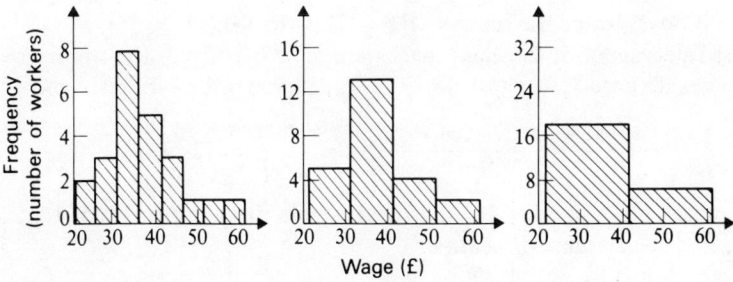

Figure D

The first frequency table in (a), in classes of 5, is probably the best picture, as there are too few bars in the other two.

We usually aim to have between 5 and 10 bars in a bar chart. This should be sufficient to 'iron out' the switchback effect of Figure 1, but still give enough to distinguish between such general shapes as those shown in Figure E.

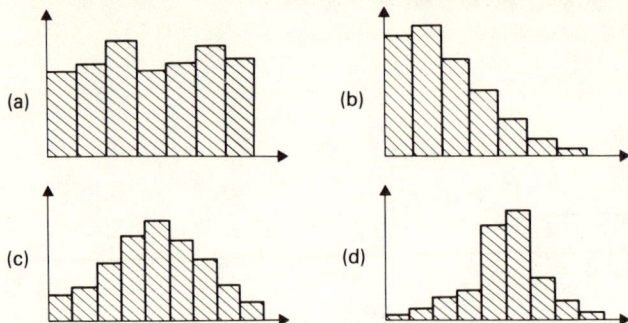

(a)

(b)

(c)

(d)

Figure E

3.2 The mean of a grouped frequency table

▷ **1**

Class	Class mark (m)	Frequency (f)	Estimated sum (m×f)
0–19	10	4	4 × 10 = 40
20–39	30	8	8 × 30 = 240
40–59	50	11	11 × 50 = 550
60–79	70	5	5 × 70 = 350
80–99	90	2	2 × 90 = 180
		30	1360

The estimated mean $= 1360 \div 30 = 45.3\ldots$ or 45 to the nearest whole number.

Exercise B

▷ **1** (a) The sum of the individual pulse rates $= 4325$.
There are 60 boys, hence the mean $= \frac{4325}{60} = 72.1$ (to 1 d.p.).

(b) Using class intervals of 5, the class marks are 52, 57, ..., 92; the frequencies (see earlier results) are 1, 4, 6, 10, 18, 9, 7, 3, 2; and so the estimated mean is:

$$\frac{(1 \times 52 + 4 \times 57 + \ldots + 2 \times 92)}{60} = \frac{4330}{60} \approx 72$$

(c) The class marks are now 55, 65, 75, 85 and 95; the frequencies are 5, 16, 27, 10, and 2; and the estimated mean is:

$$\frac{(5 \times 55 + \ldots + 2 \times 95)}{60} = \frac{4380}{60} = 73$$

But, strictly, the class marks should be $54\frac{1}{2}$, $64\frac{1}{2}$, ... and so a better estimate, using a class interval of 10, is $73 - \frac{1}{2} = 72\frac{1}{2}$. (If you are using a calculator, then it is just as easy to work out $\dfrac{(5 \times 54\frac{1}{2} + 16 \times 64\frac{1}{2} + \ldots + 2 \times 94\frac{1}{2})}{60}$ straight away.)

As might be expected, (b) gives a closer estimate to the true mean.

> **2** The class marks are 5, 15, ..., 55 (pence); the frequencies are 5, 7, 7, 13, 17, 11 (children); and so the estimated mean amount of money per child is

$$\frac{5 \times 5 + 7 \times 15 + \ldots + 11 \times 55}{60} \approx 35 \text{ pence.}$$

3 The class marks are 3, 8, 13, ..., 33, and so the estimated mean

$$= \frac{15 \times 3 + \ldots + 1 \times 33}{100} = \frac{1225}{100} \approx 12 \text{ words.}$$

3.3 Mode, median and cumulative frequency

The modal class

> **1** There is no single modal pulse rate, since rates of 70 and 72 both occur 5 times.

2 With a class interval of 5, the modal class is 70–74. With a class interval of 10, the modal class is 70–79. The first of these is the more informative, as it 'narrows down' the possible range. (The middle of the modal class is sometimes taken as an estimate of the mode, so in this example we could say the estimated mode is 72. The fact that this coincides with one of the modes of the individual frequency diagram is purely coincidental!)

The cumulative frequency curve

> **3** The frequencies are 4, 8, 11, 5, 2 and the cumulative frequencies are 4, 12, 23, 28 and 30. As a check, note that the last of the cumulative frequencies must be the same as the sum of the separate frequencies.

4 Twelve pupils scored less than 40.

5 Seven pupils scored 60 or more (as twenty-three of the thirty pupils scored less than 60).

Exercise C

> **1** (a) The cumulative frequencies are 3, 20, 61, 146, 243, 358, 459, 523, 544, 550.
>   (b) See Figure F.

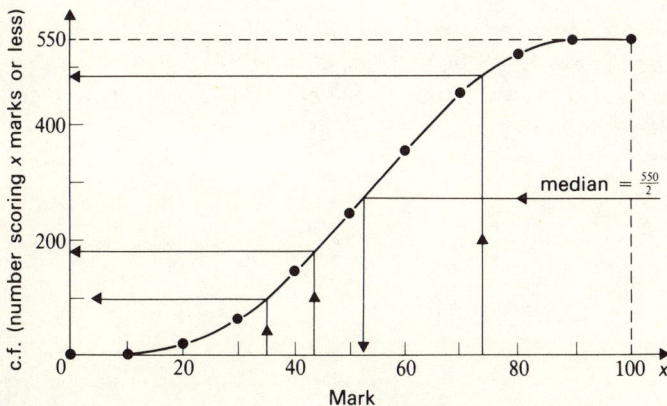

Figure F

(c) The halfway person is the $\frac{550}{2}$th or 275th. His (or her) mark, and therefore the median, is 52 (see Figure F).

The number of candidates who scored 35 or less is also marked on Figure F. Starting at 35 on the 'marks' axis, we read off that about 100 people scored 35 or less.

Starting at 74 on the 'marks' axis, about 485 people scored 74 or less, so about 65 (550 − 485) people scored 75 or more.

About 180 scored 44 or less, so about 370 people passed.

(d) 60% of 550 = 330. If 220 people failed, then the pass mark would be about 49. (220 people scored 48 or less, and so the remaining 330 scored more than 48. As a check on this, consider the following argument. There are nearly 100 people in the 41–50 class, so we could assume that about 10 people scored exactly 50, 10 scored exactly 49, etc. Altogether, 243 people scored 50 or less; i.e. 307 people scored over 50, hence about 317 scored over 49, about 327 scored *over* 48, etc., i.e. a minimum of 49.)

The class intervals were given as 41–*50*, etc., and so if we plot 243 against a mark of 50 we are implying that 243 people scored 50 *marks or less*. Compare this with questions **4** and **5** in Section 3.3.

11▷ **2** The cumulative frequencies are 7, 35, 118, 288, 387, 459, 489, 499.

From Figure G the median is about 440. As this is 7 below the correct answer, it suggests that the competitors were under-guessing – but only slightly.

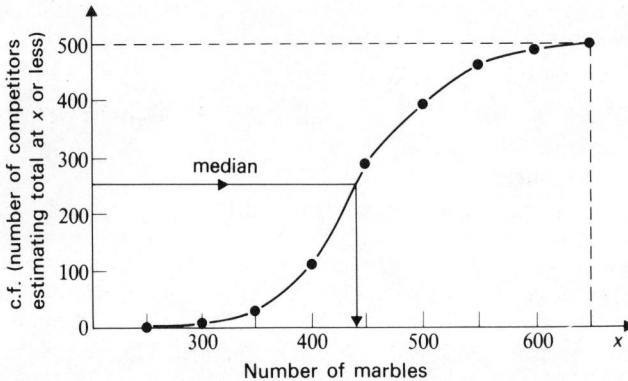

Figure G

3 The cumulative frequencies are 7, 36, 79, 95, 100.

From Figure H the median mass is about 11.346 or 11.347 grams. By calculation: median = $11.3445 + \frac{14}{13} \times 0.005$ grams ≈ 11.346 grams

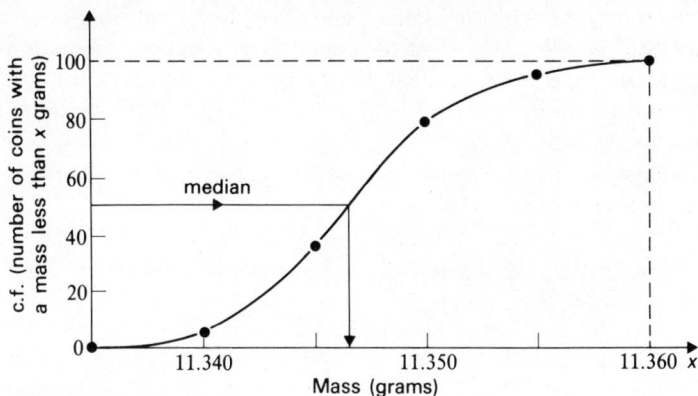

Figure H

3.4 Quartiles and inter-quartile range

Quartiles

> 1   The lower quartile is 38 or 39 marks, the upper is 65.
2   The median is the middle quartile.
3   $65 - 39 = 26$ marks.
It happens in this example that the median (52 marks) is halfway between the two quartiles. This is not always so, although it is often *approximately* halfway; especially if the frequency diagram is roughly symmetrical.
    In this example we can say that the middle half of the results are in the range 52 ± 13 marks.

Exercise D

> 1   The upper quartile is the mark of the 'halfway person in the top half', i.e. the mark of the 23rd person. From the graph (and the table in this particular case) the 23rd person scored 'less than 60', i.e. 59; and so the upper quartile is 59. Similarly the lower quartile is 31.
    The inter-quartile range (or IQR) is $59 - 31 = 28$.
    The median for this class is 45, and so the middle half scored marks in the range 45 ± 14.
    If this class is a sub-set of the population of Exercise C, question 1, then it is 'below average' (with a median mark 7 lower), but, interestingly, there is very little difference in the spreads (IQRs of 26 and 28). (If the total population were a year-group of a town's comprehensive schools, then this class of 30 might be an unstreamed class, that, for *some* reason, was below average.)
2   For Exercise C, question 2, the lower quartile is 405, the upper quartile 490, and the IQR is 85 (marbles), approximately. Half the competitors made a guess within the range 405–490.
    For Exercise C, question 3, LQ = 11.343, UQ = 11.349, IQR = 0.006 (grams). There is very little variation in the mass of a 10p piece; this is why banks are able to 'weigh' money such as a £5 bag of silver to check it.

3 (a) 125 people.

(b) An enlargement of the relevant part of the cumulative frequency diagram is shown in Figure I.

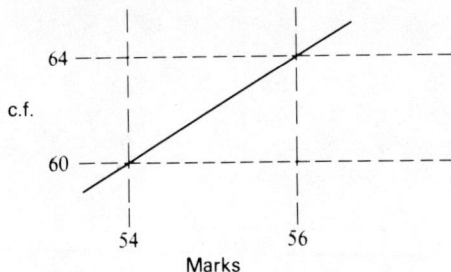

Figure I

62 people scored 55 marks or less
64 people scored 56 marks or less

Hence the 62nd person probably scored 55, and the 63rd and 64th persons 56, so that the median is about $55\frac{1}{2}$. A sensible answer would be '55 or 56'.

UQ (mark of $94\frac{1}{2}$ person) ≈ 70
LQ (mark of $31\frac{1}{2}$ person) ≈ 39
IQR $\approx 70 - 39 = 31$

(c) Since 32 people scored 39 marks or less, then 93 $(125 - 32)$ passed by scoring 40 marks or more.

4 The cumulative frequencies are 1, 3, 14, 28, 66, 113, 164, 196, 200.

Note that 14 (for example) is plotted against 65 k.p.h., as the vehicles in the '60' class could have been travelling at any speed from 55 k.p.h. up to 65 k.p.h.

(a) See Figure J. Median is 92 k.p.h. IQR $= 102 - 81 = 21$ k.p.h. (all approximate).

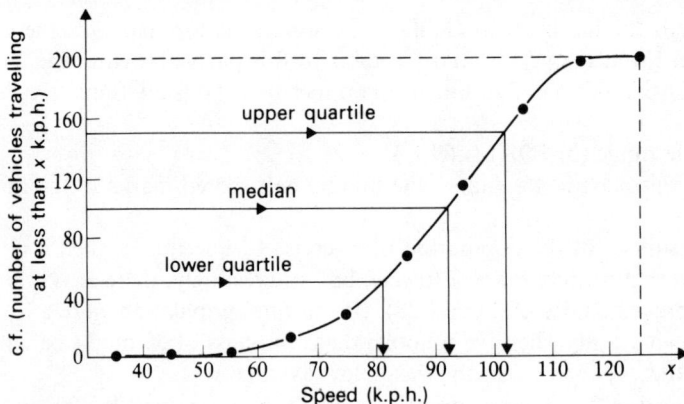

Figure J

(b) About 138–140 vehicles were travelling at speeds up to 100 k.p.h. As speed is a continuous variable (i.e. it is not measured only in whole numbers) it is unlikely that any vehicles were travelling at exactly 100 k.p.h., so the remaining 60–62 vehicles were travelling at over 100 k.p.h.; i.e. *approximately* 30%.

Post-test

1 (a)

Class	0–4	5–9	10–14	15–19	20–24	25–29	30–34	35–39	40–44	45–49
Frequency	4	4	7	8	6	7	6	8	7	1

(b) See Figure K.

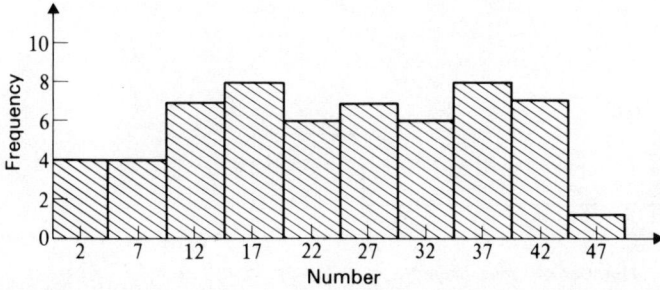

Figure K

(c) $1408 \div 58 = 24.3$ (correct to 1 d.p.)

$$\frac{(4 \times 2 + 4 \times 7 + 7 \times 12 + \ldots + 1 \times 47)}{58} = \frac{1406}{58} = 24.2 \text{ (correct to 1 d.p.)}$$

2 See Figure L.

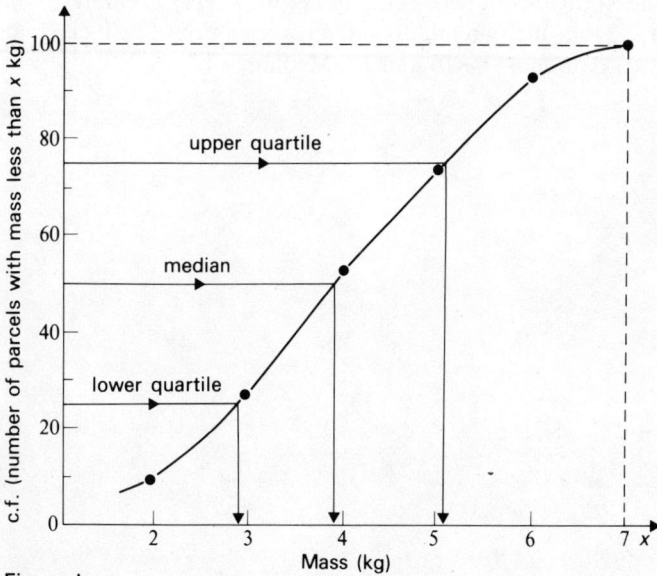

Figure L

The median is about 3.9 kg.
The IQR is 5.0(5) − 2.9 kg, i.e. between 2.1 and 2.2 kg.

61

3 (a) The mean number of appearances $= 511 \div 24 = 21.3$ (or 21).

(b)

Class	1–10	11–20	21–30	31–40	41–50
Frequency	11	2	2	2	7
Cumulative frequency	11	13	15	17	24

(The class intervals could be 0–9, 10–19, etc.)

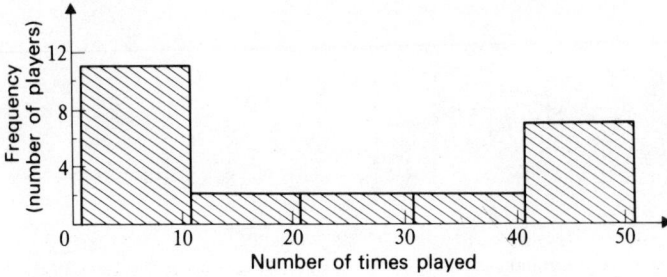

Figure M

(c) See Figure M. The mean of the grouped frequencies is:

$$\frac{(11 \times 5 + 2 \times 15 + 2 \times 25 + 2 \times 35 + 7 \times 45)}{24} = 21.7 \approx 22$$

Because of the small population (24 people) and the wide range of the variable (1 to 45 appearances) this is not necessarily going to be very accurate.

(d) From (b) we can see that the two middle numbers are in the 11–20 class, and so from the original data must be 16 and 18. Median $= 17$.

4 Further probability

Objectives

This is what you should know after studying this chapter.
(1) Whether the selection of a particular outcome from a set of outcomes is random or not.
(2) If two outcomes A and B are mutually exclusive, then $p(A$ and $B) = 0$ and $p(A$ or $B) = p(A) + p(B)$ (i.e. for mutually exclusive outcomes we add their probabilities to obtain the probability of one or other of the outcomes).
(3) How to use tree diagrams in problems with combined events.
(4) When considering the probability of outcomes associated with combined events, $p(A$ followed by $B) = p(A) \times p(B)$.

Pre-test

> 1   If the probability of a person chosen at random being left-handed is $\frac{1}{20}$, what is the probability of a person chosen at random being right-handed? Is it possible to use this result to determine the number of left-handed batsmen in a cricket eleven?

2   (a)   What is the theoretical (or expected) probability of throwing a five with an unbiased die?
Using this die, how many fives would you expect from 10 throws; 50 throws; 500 throws; 1000 throws?

      (b)   The table shows how often a five turns up when a die is thrown a certain number of times. What would you give as the probability of throwing a five with this die?
Do you think the die is biased?

No. of throws	10	50	500	1000
No. of fives	2	7	94	172

      (c)   John throws his die 1200 times. His 'success fraction' for getting a six is $\frac{461}{1200}$. What does this suggest about his die?

3   (a)   Explain why, when two unbiased coins are tossed together, the probability of throwing two heads is not $\frac{1}{3}$.

63

(b) What are the equally likely outcomes when three unbiased coins are tossed together?

(c) List the elements of the possibility space when a coin and a die are thrown together.

4.1 Random selection

When a prize draw takes place, it is a basic principle that the draw is so arranged that each ticket has an equal chance of winning. The common method used is for all tickets to be placed into a hat or drum and thoroughly shaken. Then someone, without looking into the hat or drum, selects the winning ticket. In this process each ticket has an equal chance of being selected.

This is called a random process. Another method sometimes used is to employ an electronic computer to print out a list of random numbers; this is very useful when there are very many entrants and many prizes to be won. You are probably familiar with the use of Ernie to draw the Premium Bonds winning numbers each week.

If nine people out of a group of ten stated that they 'cannot tell Stork margarine from butter', it would not be a justifiable conclusion that 90% of the population of this country could not do so. If, however, about 9000 people out of a sample of 10000 (drawn from different parts of the country) had the same difficulty in distinguishing between Stork and butter, then the conclusion is much more likely to be justified. Predictions made from a large sample, *chosen properly at random*, are more valid than those from a small sample. (In fact, it can be shown that – provided the sample really is random – 2000 is a large enough number to give 'good' results.)

Look at the letters in the first paragraph of this section. They appear to be randomly distributed: if you closed your eyes and placed a pinpoint on the page could you say what letter would be closest to it? Are all of the letters of the alphabet equally likely, or is, say, E more likely than G?

The table gives the results of a frequency count of the letters in the paragraph.

Letter	Frequency	Prob-ability	Letter	Frequency	Prob-ability	Letter	Frequency	Prob-ability
A	28	0·09	J	0	0.00	S	18	0.06
B	3	0.01	K	7	0.02	T	31	0.10
C	18	0.06	L	11	0.04	U	7	0.02
D	10	0.03	M	6	0.02	V	0	0.00
E	35	0.12	N	23	0.08	W	6	0.02
F	3	0.01	O	20	0.07	X	0	0.00
G	6	0.02	P	6	0.02	Y	1	—
H	22	0.07	Q	2	0.01	Z	1	—
I	23	0.08	R	13	0.04	Totals	300	0.99

(The probability of A is found from $\frac{28}{300}$; etc. The sum of the *exact* probabilities should be 1. The 0.01 has been 'lost' either in rounding, or could be regarded as the combined probability of Y and Z.)

Are these results typical of written English as a whole?

Question **3** of the next exercise will consider this. The point at the moment is that the letters of the alphabet do not occur with equal frequency, so if you have to 'choose a letter at random' it is not good enough to pick out a letter from a page with a pin!

Exercise A

> **1** You are required to select at random one number from the first twelve counting numbers 1, 2, ..., 12. Which of the following methods are valid?

   (a) Draw a card from a pack of playing cards. (Take the Jack to represent 11, the Queen 12, and draw again if it is a King.)

   (b) Throw a pair of dice, and add together the two numbers.

   (c) Number the faces of a regular unbiased dodecahedron (twelve-faced solid) 1 to 12, and throw as a die.

   (d) Use a correspondence such as A or M = 1, B or N = 2, ... (ignore Q and Z) and pick out the 3rd letter of the 2nd word of the 5th line of the 10th page of a book.

   (e) Close your eyes and think of a number less than 13.

> **2** Figure 1 shows two fixed discs, each fitted with a rotating arrow. If the arrow is spun freely, state the probability that the arrow in (a) will stop on the unshaded

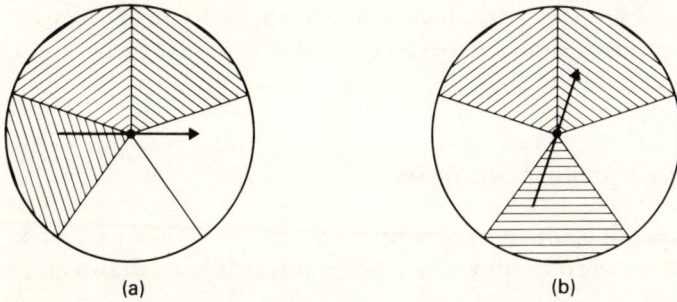

(a)   (b)

Figure 1

part. If the arrow in (b) is spun freely, is it more or less likely to stop on the unshaded part?

3 Make a frequency count of the letters in the paragraph below, and compile a table similar to the one above. Compare the results.

'Statistics is often called the science of large numbers. We cannot say definitely what will happen in a particular event; but we can use statistics to make a prediction about the outcomes of a large number of events.'

4.2 Expected probability

In Chapter 2 we considered two different ways of obtaining the probability of a certain outcome for a given event: experimental and theoretical. In using these ideas we must be careful that the procedures we use are valid. It may be helpful to state again the conditions to be observed in each case:

(1)   When using the experimental method, we should carry out a large number of trials.

(2)   When working out the expected probability by theoretical considerations, we must ensure that all the outcomes are equally likely.

Exercise B

1   Find the expected probability of each of the following results.
(a)   A number less than 3 when an unbiased die is thrown.
(b)   A total less than 3 when two unbiased dice are thrown together.
(c)   A diamond when a card is drawn from a normal pack of 52 playing cards.
(d)   Three heads when three coins are tossed together.
(e)   A red marble when a marble is drawn at random from a bag containing 3 red, 4 blue and 2 white marbles.

2   (a)   A die is thrown 1000 times; about how many times would you expect an odd number to turn up?
(b)   What would you say about a certain coin that, when tossed 10 times gave 1 tail; when tossed 100 times gave 23 tails; when tossed 1000 times gave 191 tails; and when tossed 5000 times gave 1017 tails?

About how many heads would you expect to get from 300 tosses of this particular coin?

3   You open a telephone directory at random, and find that there are 750 numbers on the double page. How many would you expect to end in 3? Would you expect about the same number to begin with 3?

4.3   Mutually exclusive outcomes

Consider the following situation. A card is drawn at random from a pack of 52 playing cards. What is the probability of the following cards being drawn?
(a)   A heart or a five.
(b)   A heart and a five.
(c)   A picture card or a five.
(d)   A picture card and a five.

For each of these trials there are 52 (equally likely) possible outcomes, represented by the 52 different cards in the pack. Hence $n(\mathscr{E}) = 52$.

Let H be the set of outcomes that produce a heart,
P the set that produces a picture card,
F the set that produces a five.

1   Write down the values of $n(H)$, $n(P)$, $n(F)$.

2   Draw two Venn diagrams to show the relation between F and H; and the relation between F and P; and write down the values of $n(F \cup H)$, $n(F \cap H)$, $n(F \cup P)$, $n(F \cap P)$.

3   Hence write down as fractions the values of $p(F)$, $p(H)$, $p(P)$, $p(a)$, $p(b)$, $p(c)$, $p(d)$.

You will have seen that there is a difference between the pair of situations (a) and (b), and the pair (c) and (d). This is shown up in the Venn diagrams.

In situations (c) and (d) we say that the outcomes F and P are *mutually exclusive*; i.e. they cannot both occur at the same time. Hence $p(F \text{ and } P) = 0$.

As a consequence, if two outcomes are mutually exclusive the probability of one or other occurring is the sum of their separate probabilities; e.g.

$$p(F \text{ or } P) = p(F) + p(P)$$

Exercise C

> 1   In a group of 30 teenagers, 22 like popular music, 12 like classical music, and 2 like neither type of music.
  (a) Draw a Venn diagram to illustrate this information.
  (b) If a member of the group is chosen at random, find the probability that he or she likes popular music only; classical music; some type of music; both popular and classical music.

2   An unbiased die is thrown once.
  (a) State the probability that the outcome is an even prime number; an odd prime number; an even square number.
  (b) Name two outcomes for this trial that are mutually exclusive.

4.4 Possibility spaces for combined events

Question **2** of Exercise A was concerned with the probability of an arrow stopping in the shaded or unshaded areas of a disc. The diagram is repeated in Figure 2.

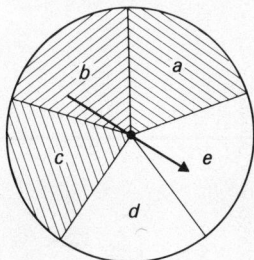

Figure 2

Let S be a spin for which the arrow stops over the shaded part, and U a spin for which it stops over the unshaded part. For a single spin, $p(S) = \frac{3}{5}$, and $p(U) = \frac{2}{5}$.

If we spin the arrow twice, what is the probability of SS? (That is, what is the probability of the arrow stopping over the shaded part twice in succession?)

The possibility space (i.e. the possible outcomes) is shown in the table.

1st spin	2nd spin
S	S
S	U
U	S
U	U

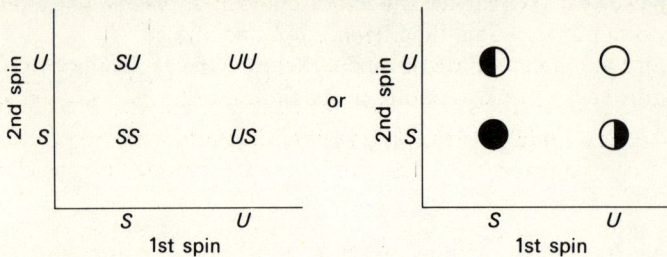

Figure 3

Figure 3 shows the possibility space diagrammatically.

We use the same procedure as in finding the probabilities when two coins were tossed twice – but in this case, since $p(S) \neq p(U)$, the four possible outcomes are *not* equally likely.

One method of overcoming this difficulty is suggested by the redrawn diagram of the disc (Figure 2). Let there be *five* possible outcomes for a single spin of the arrow, corresponding to the arrow stopping over the five sectors a, b, c, d and e. *These outcomes are equally likely.* (And if we let $S = \{a, b, c\}$ and $U = \{d, e\}$, then $n(S) = 3$, and $p(S) = \frac{3}{5}$, etc. as before.) For two spins we can now draw the possibility space as in Figure 4. This consists of 25 *equally likely* possible outcomes; giving the results $p(SS) = \frac{9}{25}$, $p(SU) = p(US) = \frac{6}{25}$ and $p(UU) = \frac{4}{25}$.

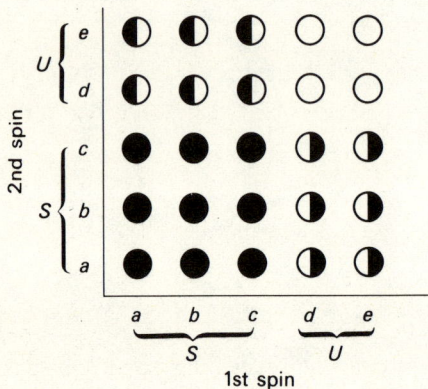

Figure 4

Figure 4 may remind you of diagrams used to illustrate the multiplication of fractions; and it leads to the result $p(A$ followed by $B) = p(A) \times p(B)$.

Exercise D

▷ 1   Draw a possibility space diagram to show the outcomes when two unbiased dice are thrown together, and find the following probabilities.

(a)   The probability of the sum on the two dice being greater than 9.

(b)   The probability of the product of the numbers on the two dice being 12.

(c)   The probability of the difference between the numbers on the two dice being less than two.

(You have already done a question like this in Chapter 2: Exercise D, question **2**.)

7

▷ **2**   A mini-pack of cards consists of the Jack, Queen and King of Hearts, and the Jack, Queen and King of Spades. Draw the possibility space diagram when two cards are drawn out of this pack. (Remember that, this time, the same card cannot be chosen twice.)

State the probability of the two cards being (a) two kings; (b) a red and a black picture card; (c) a male of one colour and a female of the other colour.

8

4.5   Tree diagrams

Listing or drawing a possibility space can be tedious (consider the possibility space when two cards are drawn from a complete pack of 52 cards!), difficult for three successive events (the diagram now needs to be three-dimensional), and impossible diagrammatically for four or more events.

An easier way is to draw a *tree diagram*. This corresponds to the simpler form of possibility space; but it is drawn in the form of a branching tree, with the probabilities of the *individual* outcomes on the branches of the trees.

The tree diagram for two spins of the arrow (see the beginning of Section 4.4) is shown in Figure 5. Note that the possible outcomes for the second spin are repeated for each outcome of the first spin.

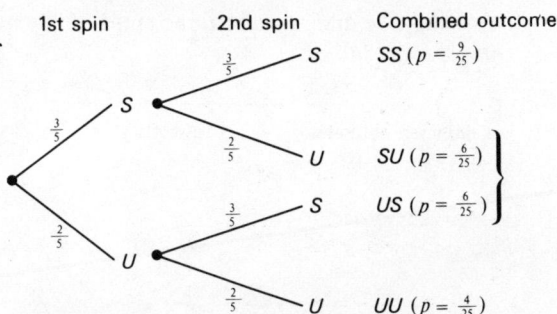

Figure 5

▷ **1**   A bag contains 7 red marbles and 3 blue marbles. One marble is taken out at random, its colour is noted, and it is then returned to the bag; and then another marble is taken out and its colour noted. Complete the tree diagram in Figure 6 to show the probabilities of the various colour combinations.

9

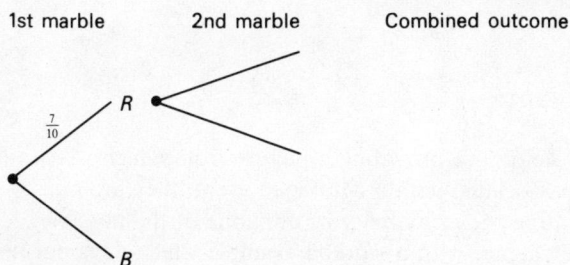

Figure 6

69

▷ **2** If, in the previous example, the first marble is not put back, then the second stage of the tree diagram has to be modified. If the first marble was red, then the probability of the second marble being red is $\frac{6}{9}$ (of the remaining nine marbles, six are red); but if the first marble is blue, then the probability of the second marble being red is now $\frac{7}{9}$ (as seven of the nine marbles still in the bag are red). Complete Figure 7.

1st marble   2nd marble   Combined outcome

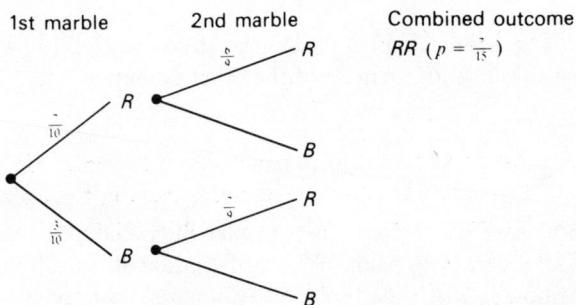

Figure 7

Exercise E

▷ **1** Draw the tree diagram for three spins of the arrow in Figure 2, and use it to complete the table.

Combined outcome	Probability
SSS	
SSU (in any order)	
SUU (in any order)	
UUU	

2 A sub-committee of three is to be chosen at random from seven men and three ladies. Find the probability that the members of the sub-committee will not all be of the same sex.

4.6 More combined events

Whilst tree diagrams are very helpful in providing a picture from which we can find the probabilities of outcomes associated with a combined event, they are not essential, particularly if we require the probability of only one of the possible outcomes. We shall finish this chapter with a worked example which does not use a possibility space or a tree diagram; and an exercise which you should try to set out similarly.

70

A housewife buys a dozen eggs, but unknown to her two of them are bad. She decides to make an omelette with three of them. What is the probability that she chooses three good eggs? What is the probability that she chooses one bad egg?

Let G stand for the choice of a good egg, and B for a bad egg. For the first egg, $p(G) = \frac{10}{12} = \frac{5}{6}$. If the first egg was good, there are 9 good eggs in the remaining 11, and so for the second egg $p(G) = \frac{9}{11}$. Similarly, if the first two eggs are good, for the third egg $p(G) = \frac{8}{10} = \frac{4}{5}$.

Hence the probability of all three eggs being good is:

$$\frac{5}{6} \times \frac{9}{11} \times \frac{4}{5} = \frac{1 \times 3 \times 2}{1 \times 11 \times 1} = \frac{6}{11}$$

The probability that she uses two good eggs and one bad egg in the order GGB is (before simplifying) $\frac{10}{12} \times \frac{9}{11} \times \frac{2}{10}$. The probability that she chooses them in the order GBG is $\frac{10}{12} \times \frac{2}{11} \times \frac{9}{10}$; and in the order BGG is $\frac{2}{12} \times \frac{10}{11} \times \frac{9}{10}$. (Note that these three probabilities are all the same.) Therefore:

$$p(GGB \text{ in some order}) = 3 \times \frac{10 \times 9 \times 2}{12 \times 11 \times 10} = \frac{9}{22}$$

Exercise F

1 Three dice are thrown. Find the probability that the sum of the numbers showing is greater than 15.

2 Two cards are dealt at random from a pack of 52 playing cards (i.e. the first card is not replaced).
   Find the probability that they are the following.
   (a) Both diamonds.
   (b) Both of the same suit.
   (c) An ace and a picture card.

3 In recent years the probability that a baby will be a boy has been 0.514. On this assumption, calculate the probabilities that a family with two children will be as follows.
   (a) Both boys.
   (b) Both of the same sex.
   (c) Of opposite sex.
   Does the answer to (c) depend upon whether the boy or the girl is the older child?

Summary

(1) If an element is to be 'drawn at random' from a population, then the method used must be such that each element of that population has an equal chance of being chosen.

(2) If we have an event for which two of the possible outcomes are represented by the sets A and B, and if $\mathscr{E}$ is the set of all the *equally likely* possible outcomes, then:

$$p(A \text{ and } B) = \frac{n(A \cap B)}{n(\mathscr{E})}$$

$$p(A \text{ or } B) = \frac{n(A \cup B)}{n(\mathscr{E})}$$

(3) When the sets A and B are mutually exclusive (i.e. the two outcomes cannot both occur at the same time), then:

$$p(A \text{ and } B) = 0 \quad (\text{since } n(A \cap B) = 0)$$
$$p(A \text{ or } B) = p(A) + p(B) \quad (\text{since } n(A \cup B) = n(A) + n(B))$$

For example, for a single throw of a die, $\mathscr{E} = \{1, 2, 3, 4, 5, 6\}$. If $A = \{1, 4\}$ and $B = \{6\}$, then A and B are mutually exclusive outcomes.

(4) *Combined events*

If we have a series of events E_1, E_2, E_3, ..., and we are interested in some associated outcomes A_1, A_2, A_3, ... for which the probabilities are $p(A_1)$, etc., then the probability of the outcome $A_1 A_2 A_3 \ldots$, in that order, is

$$p(A_1 A_2 A_3 \ldots) = p(A_1) \times p(A_2) \times p(A_3) \times \ldots$$

(5) *Possibility spaces*

The possible outcomes of two combined events can be illustrated by a possibility space diagram, showing all the possible outcomes. If these are equally likely, then the diagram can be used to work out easily the probability of any particular (set of) outcomes.

(6) *Tree diagrams* can also be used to illustrate combined events, and to obtain the probabilities of various outcomes. They can often be used more easily and more widely than possibility space diagrams.

Example

Two cards are drawn from a pack of 52 playing cards. What is the probability that one is a picture card, and the other is not (if the ace is *not* counted as a picture)?

Two events are involved here. E_1 is 'a card is drawn from a pack of 52 cards', and E_2 is 'a card is drawn from a pack of 51 cards'. Associated with E_1 are the outcomes A_1: 'a picture card is drawn', and B_1: 'a non-picture card is drawn'; and similarly with E_2 the outcomes A_2 and B_2 are associated. (Note that A_1 and B_1 are mutually exclusive, as are A_2 and B_2.)

For this example we are interested in the combined outcomes $A_1 B_2$ (i.e. A_1 followed by B_2) and $B_1 A_2$. $p(A_1) = \frac{12}{52} = \frac{3}{13}$, and $p(B_1) = \frac{40}{52} = \frac{10}{13}$; but the values of $p(A_2)$ and $p(B_2)$ depend on the outcome of the first event. If the outcome of E_1 was A_1 then $p(A_2) = \frac{11}{51}$ and $p(B_2) = \frac{40}{51}$; but if the outcome was B_1, then $p(A_2) = \frac{12}{51}$ and $p(B_2) = \frac{39}{51}$.

$$\text{Thus } p(A_1 B_2) = \frac{3}{13} \times \frac{40}{51}, \quad \text{and} \quad p(B_1 A_2) = \frac{10}{13} \times \frac{12}{51}.$$

Both of these are equal to $\frac{40}{221}$ ($A_1 B_2$ and $B_1 A_2$ are equally likely) and the probability we require is $2 \times \frac{40}{221} = \frac{80}{221}$. ($A_1 B_2$ and $B_1 A_2$ are mutually exclusive.)

Figure 8 shows this presented as a possibility space and as a tree diagram. Note that in the possibility space it is not possible to have outcomes on the diagonal, as the same card cannot be drawn twice.

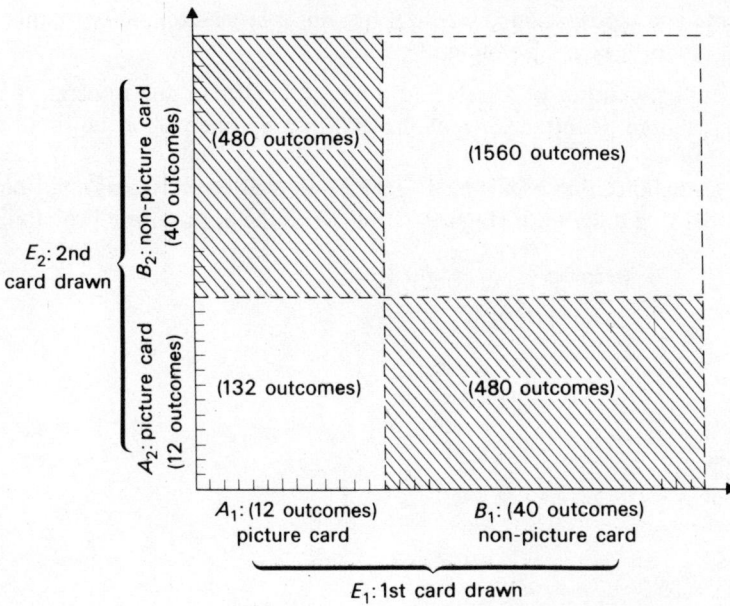

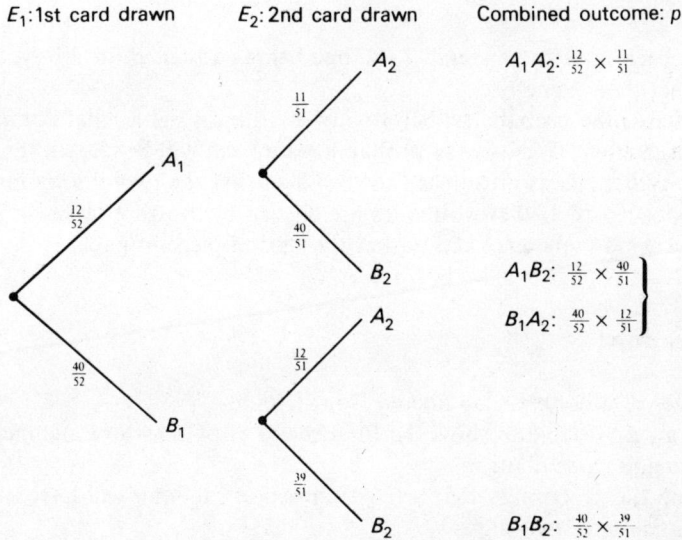

Figure 8

E_1: 1st card drawn	E_2: 2nd card drawn	Combined outcome: p
	A_2	$A_1 A_2$: $\frac{12}{52} \times \frac{11}{51}$
A_1		
	B_2	$A_1 B_2$: $\frac{12}{52} \times \frac{40}{51}$
	A_2	$B_1 A_2$: $\frac{40}{52} \times \frac{12}{51}$
B_1		
	B_2	$B_1 B_2$: $\frac{40}{52} \times \frac{39}{51}$

Post-test

> 1 Ladies' blouses are made in three sizes – small, medium and large. In a group of 24 ladies it is found that 5 can wear the large size (of these 3 can also wear the medium size, but none the small size); 18 altogether can wear the medium size; and 7 altogether can wear the small size. (All the ladies can wear at least one of the sizes.)

(a) Draw a Venn diagram to illustrate this information.

(b) If one of these ladies is chosen at random, what is the probability that she can wear a large size blouse; a medium size, but not a large size, blouse; either a medium size or a large size blouse?

2 From a youth club committee of 5 girls and 6 boys a chairman and a secretary are to be chosen at random, with the obvious proviso that no one person can hold both offices.

(a) State the probability that a particular girl, Anne, will be chosen as chairman.

(b) Complete the tree diagram in Figure 9, and write the appropriate probability

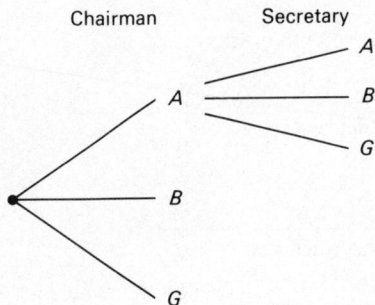

Chairman   Secretary

Figure 9

on each branch. (*A* stands for Anne being chosen, *B* for a boy, and *G* for a girl other than Anne.)

(c) What is the probability, before any selection is made, that Anne will be chosen as secretary? What is the probability that she will be chosen for neither office?

(d) A further rule is introduced to the effect that the two officers must be of opposite sex. If the two names are chosen by drawing pieces of paper out of a hat, what is the probability that the second piece of paper is invalid?

13

Assignment

1 The pages of a book are numbered from 1 to 300.

(a) Draw up a table to show the frequencies of pages whose numbers have one, two and three digits.

(b) Find the probability that a page, chosen at random, will have a number consisting of two digits.

(c) A page is chosen at random from those having three-digit numbers. What is the probability that just one of the digits will be 0?

2 The country lanes shown in Figure 10 have no signposts; and a motorist exploring the area has no map, so at each junction he has to guess which way to go. If he *can* go straight on, the probability that he will do so is $\frac{2}{5}$; where both left and right turns are possible they are equally likely to be taken. The motorist never turns back!

(a) Find the probabilities that he turns left at X if he came from P, and if he came from Q.

(b) What are the probabilities that he will travel on to Q from Y if he came from Q; R; S?

74

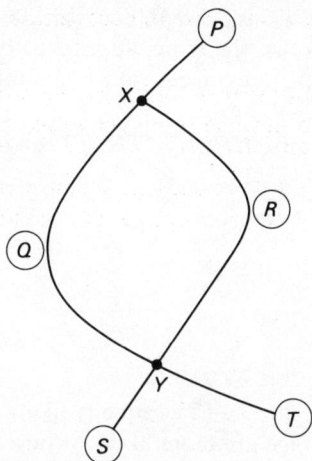

Figure 10

(c) The motorist sets out from P. Find the probabilities of reaching T by each of the following routes: PRT; one of the two sensible routes; $PRQRT$ (assuming that he does not learn from his mistakes).

3 An examination question has five parts, each requiring a single answer which is either right or wrong. A correct answer scores $^+2$ marks, a wrong answer $^-1$ mark, and no attempt scores 0 marks.

(a) What are the maximum and minimum marks for the question?

(b) List the possible ways that 0 can be scored for the question.

(c) A candidate attempts all parts, and is equally likely to be right or wrong in each part. State the probabilities of obtaining the following scores:
10; 9; 5; 0; a negative score

Answers

Pre-test

> 1 $p(\text{right-handed}) = 1 - \frac{1}{20} = \frac{19}{20}$.
No; 11 is far too small a sample to be representative.

2 (a) $p(5) = \frac{1}{6}$. For 10 throws, $\frac{1}{6} \times 10 \approx 2$; for 50 throws, $\frac{1}{6} \times 50 \approx 8$; for 500 throws, $\frac{1}{6} \times 500 \approx 83$; and for 1000 throws, $\frac{1}{6} \times 1000 \approx 167$.

(b)

No. of throws	10	50	500	1000
No. of fives	2	7	94	172
Probability	$\frac{2}{10} = 0.20$	$\frac{7}{50} = 0.14$	$\frac{94}{500} \approx 0.19$	$\frac{172}{1000} \approx 0.17$

The first two samples are too small for the results to be meaningful; and so the answer would be 'about 0.17 to 0.18'.

The expected probability is $\frac{1}{6}$ (≈ 0.17), and so this die is unlikely to be biased.

75

(c) For an unbiased die the expected probability of a 6 is $\frac{1}{6}$, or $\frac{200}{1200}$ (i.e. in 1200 throws you would expect about 200 sixes). This die, therefore, appears to be very heavily loaded in favour of 6, or maybe it has two faces with the number six on!

3 (a) The *equally likely* possibilities are, as ordered pairs, *HH, HT, TH, TT*, giving $p(HH) = \frac{1}{4}$.

(b) *HHH, HHT, HTH, HTT, THH, THT, TTH, TTT*.

(c) *H*1, *H*2, *H*3, *H*4, *H*5, *H*6, *T*1, *T*2, *T*3, *T*4, *T*5, *T*6.

4.1 Random selection

Exercise A

1 Parts (a) and (c) would be valid, as they both give 1, 2, ..., 12 as equally likely outcomes. In (b) the numbers in the middle of the range are more likely to turn up (see Chapter 2, Exercise D, question **2**). In (d) some letters of the alphabet are more likely to turn up than others (see Chapter 1, Exercise B, question **4**). For (e) people in practice tend to choose a number in the middle of the range, and in particular 7.

2 (a) $\frac{2}{5}$ (b) The same, $\frac{2}{5}$

3

Letter	Frequency	Prob-ability	Letter	Frequency	Prob-ability	Letter	Frequency	Prob-ability
A	17	0.10	J	0	0.00	S	13	0.07
B	4	0.02	K	1	0.01	T	20	0.11
C	10	0.06	L	8	0.05	U	7	0.04
D	3	0.02	M	4	0.02	V	2	0.01
E	23	0.13	N	13	0.07	W	4	0.02
F	5	0.03	O	10	0.06	X	0	0.00
G	2	0.01	P	4	0.02	Y	2	0.01
H	4	0.02	Q	0	0.00	Z	0	0.00
I	13	0.07	R	7	0.04	Totals	176	0.99

Even with small samples such as these, the pattern is very similar (compare the probabilities, not the frequencies, as the passages are of different length!). H appears to be the only one that differs a lot.

Compare also with the 'order of frequency' in general in written English:
ETOAN IRSHD LCWUM FYGPB VKXQJZ

4.2 Expected probability

Exercise B

1 (a) The equally likely outcomes are 1, 2, 3, 4, 5, 6. Two of these, 1 and 2, give the required outcome: $p = \frac{1}{3}$.

(b) There are now 36 equally likely outcomes (Chapter 2, Exercise D, question **2**) of which only one, $1+1$, gives a total less than 3; $p = \frac{1}{36}$.

(c) $\frac{13}{52} = \frac{1}{4}$.

76

(d) $\frac{1}{8}$ (Chapter 2, Exercise D, question **4**).

(e) $\frac{3}{9} = \frac{1}{3}$.

2 (a) The expected probability for an odd number is $\frac{3}{6}$ or $\frac{1}{2}$; hence we should expect about $\frac{1}{2} \times 1000$ or 500 odd numbers.

(b) The experimental probabilities, as decimals, are, for 10 throws, 0.1 (sample too small to be significant); for 100 throws, 0.23; for 1000 throws, 0.19; and for 5000 throws, 0.20.

This coin is very heavily biased in favour of heads, and seems to give $p(T) = 0.2$, $p(H) = 0.8$. Hence from 300 tosses we should expect about $0.8 \times 300 = 240$ heads.

3 For the *end* digit we should expect each digit to be equally likely, and thus there should be about 75 ending in 3. However, this is not true for the first digit, which is often part of a dialling code, and so the results would differ considerably from area to area.

4.3 Mutually exclusive outcomes

> 1 $n(H) = 13$, $n(P) = 12$, $n(F) = 4$.

2 See Figure A. $n(F \cup H) = 16$, $n(F \cap H) = 1$, $n(F \cup P) = 16$, $n(F \cap P) = 0$.

3 $p(F) = \frac{1}{13}$, $p(H) = \frac{1}{4}$, $p(P) = \frac{3}{13}$, $p(a) = \frac{4}{13}$, $p(b) = \frac{1}{52}$, $p(c) = \frac{4}{13}$, $p(d) = 0$.

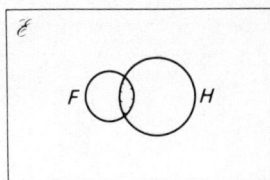

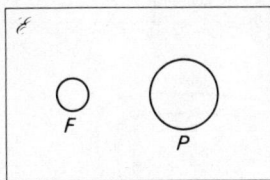

Figure A

Exercise C

> 1 (a) See Figure B.

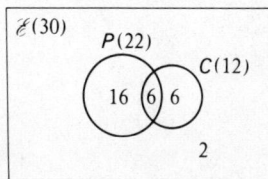

Figure B

(b) $p(\text{only popular}) = \frac{16}{30} = \frac{8}{15}$; $p(\text{classical}) = \frac{12}{30} = \frac{2}{5}$; $p(\text{some music}) = \frac{28}{30} = \frac{14}{15}$; $p(\text{both popular and classical}) = \frac{6}{30} = \frac{1}{5}$.

2 (a) For an even prime number, the outcome must be 2: $p = \frac{1}{6}$; for an odd prime number, the outcome could be 3 or 5: $p = \frac{2}{6} = \frac{1}{3}$; for an even square number, the outcome must be 4: $p = \frac{1}{6}$.

(b) 'An even number' and 'an odd number' are mutually exclusive. 'A prime number' and 'a square number' are mutually exclusive.

4.4 Possibility spaces for combined events

Exercise D

⊳ 1   (a)   See set A in Figure C. $p = \frac{6}{36} = \frac{1}{6}$.
       (b)   See set B in Figure C. $p = \frac{4}{36} = \frac{1}{9}$.
       (c)   See set C in Figure C. $p = \frac{16}{36} = \frac{4}{9}$.

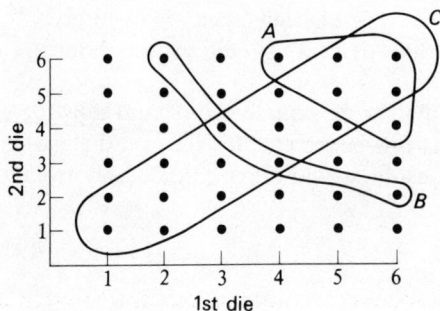

Figure C

⊳ 2   (a)   See set P in Figure D. $p = \frac{2}{30} = \frac{1}{15}$.
       (b)   See set Q in Figure D. $p = \frac{18}{30} = \frac{3}{5}$.
       (c)   See set R in Figure D. $p = \frac{8}{30} = \frac{4}{15}$.

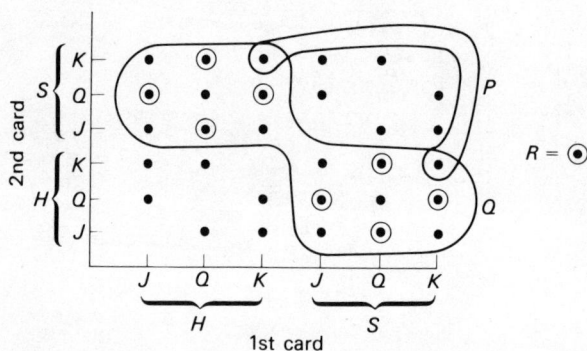

Figure D

4.5 Tree diagrams

⊳ 1   See Figure E.

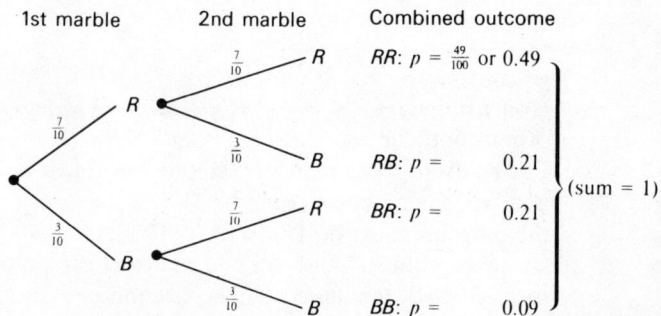

Figure E

78

> **2** See Figure F.

1st marble   2nd marble   Combined outcome

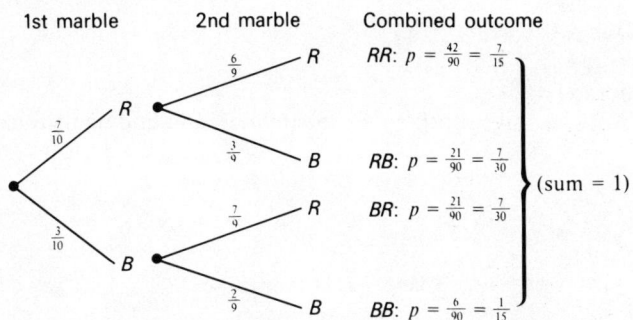

RR: $p = \frac{42}{90} = \frac{7}{15}$

RB: $p = \frac{21}{90} = \frac{7}{30}$

BR: $p = \frac{21}{90} = \frac{7}{30}$ (sum = 1)

BB: $p = \frac{6}{90} = \frac{1}{15}$

Figure F

Exercise E

> **1** See Figure G. The probabilities of the combined outcomes have been expressed as decimals for convenience.

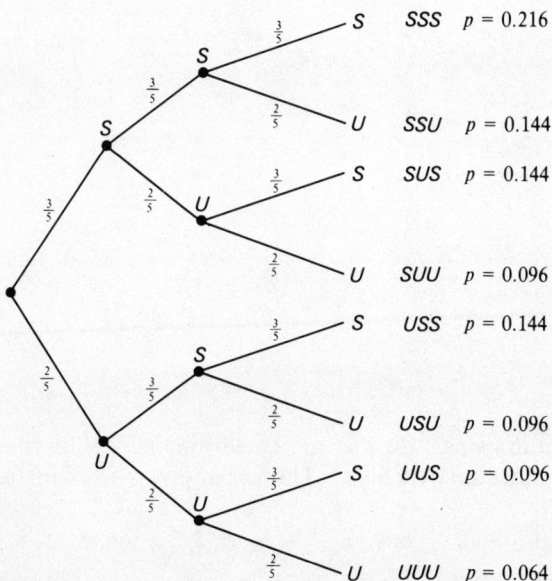

SSS $p = 0.216$

SSU $p = 0.144$

SUS $p = 0.144$

SUU $p = 0.096$

USS $p = 0.144$

USU $p = 0.096$

UUS $p = 0.096$

UUU $p = 0.064$

Figure G

Combined outcome	Probability
SSS	0.216
SSU (in any order)	$3 \times 0.144 = 0.432$
SUU (in any order	$3 \times 0.096 = 0.288$
UUU	0.064

2 See Figure H.

$p(\text{all ladies}) = p(LLL) = \frac{1}{120}.$

$p(\text{all men}) = p(MMM) = \frac{7}{24} \text{ (or } \frac{35}{120}).$

Hence $p(\text{all same sex}) = \frac{1}{120} + \frac{35}{120} = \frac{36}{120} = \frac{3}{10}.$

Hence $p(\text{not all the same sex}) = 1 - \frac{3}{10} = \frac{7}{10}.$

(You can check this by working out the other six probabilities, and finding their sum!)

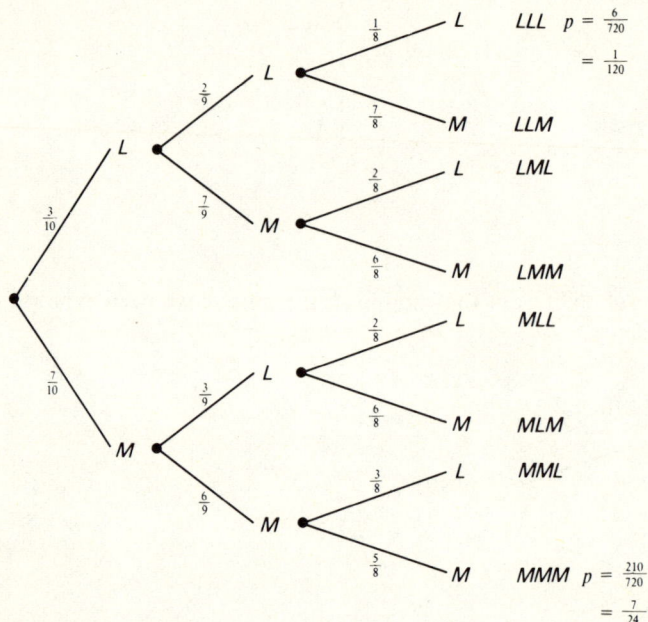

Tree diagram (Figure H):

First branch: L with $\frac{3}{10}$, M with $\frac{7}{10}$.

From L: L with $\frac{2}{9}$, M with $\frac{7}{9}$.

From M: L with $\frac{3}{9}$, M with $\frac{6}{9}$.

- $L \to L \to L$ ($\frac{1}{8}$): LLL $\quad p = \frac{6}{720} = \frac{1}{120}$
- $L \to L \to M$ ($\frac{7}{8}$): LLM
- $L \to M \to L$ ($\frac{2}{8}$): LML
- $L \to M \to M$ ($\frac{6}{8}$): LMM
- $M \to L \to L$ ($\frac{2}{8}$): MLL
- $M \to L \to M$ ($\frac{6}{8}$): MLM
- $M \to M \to L$ ($\frac{3}{8}$): MML
- $M \to M \to M$ ($\frac{5}{8}$): MMM $\quad p = \frac{210}{720} = \frac{7}{24}$

Figure H

4.6 More combined events

Exercise F

1 This is best done by listing (not drawing!) the relevant part of the possibility space.

There are 216 equally likely outcomes ($6 \times 6 \times 6$). Those that give a score of 16 or over are:

6, 6, 6   6, 6, 5   6, 5, 6   5, 6, 6   6, 6, 4   6, 4, 6   4, 6, 6   6, 5, 5   5, 6, 5   5, 5, 6

Hence probability is $\frac{10}{216} = \frac{5}{108}$.

2 (a) $\frac{13}{52} \times \frac{12}{51} = \frac{1}{17}.$

(b) $p(\text{both spades}) = p(\text{both diamonds})$, etc.

Hence $p(\text{both the same suit}) = 4 \times \frac{1}{17} = \frac{4}{17}.$

(c) In the order [ace, picture] the probability is $\frac{4}{52} \times \frac{12}{51}$, and in the order [picture, ace] it is $\frac{12}{52} \times \frac{4}{51}$. Hence $p(\text{ace and picture}) = 2 \times \frac{4}{52} \times \frac{12}{51} = \frac{8}{221}.$

3 (a) $0.514 \times 0.514 \approx 0.264.$

(b) $p(\text{both girls}) = 0.486 \times 0.486 \approx 0.236.$

$p(\text{both the same sex}) \approx 0.264 + 0.236 \approx 0.5.$

(Note that this last result is approximate. A more exact result is 0.500392.)

(c) Either $2 \times 0.514 \times 0.486 = 0.499608 \approx 0.5$ (it doesn't matter whether the boy or the girl comes first); or $1 - 0.500392 = 0.499608 \approx 0.5.$

In this example we are assuming that we are dealing with a very large population, and so the number of equally likely possibilities (which has to be expressed as a probability, as we don't know the actual size of the population) remains the same.

Post-test

> 1   (a)   See Figure I.

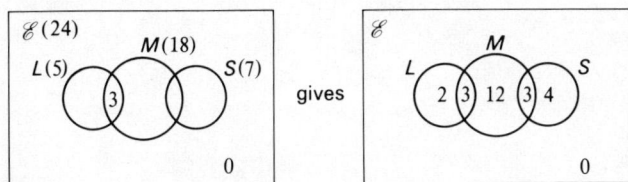

Figure I

(b) $p(\text{large}) = \frac{5}{24}$; $p(\text{medium but not large}) = \frac{(18-3)}{24} = \frac{5}{8}$;

$p(\text{medium and large}) = \frac{3}{24} = \frac{1}{8}$.

2   (a) $\frac{1}{11}$.

(b)   See Figure J.

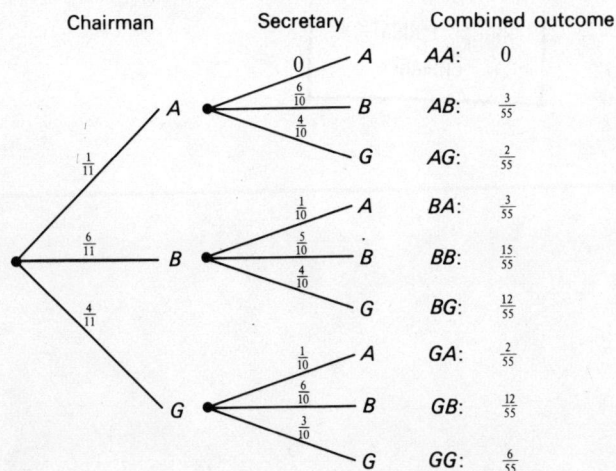

Figure J

(c) $p(A \text{ is secretary})$ is the same as $p(A \text{ is chairman}) = \frac{1}{11}$. As these are mutually exclusive, $p(A \text{ is an officer}) = \frac{2}{11}$. Hence $p(A \text{ is neither officer}) = 1 - \frac{2}{11} = \frac{9}{11}$.

(d)   The outcomes AG, BB, GA and GG are invalid. As these are mutually exclusive outcomes, $p(\text{second piece of paper invalid}) = \frac{2}{55} + \frac{15}{55} + \frac{2}{55} + \frac{6}{55} = \frac{25}{55} = \frac{5}{11}$.

Published by the Press Syndicate of the University of Cambridge
The Pitt Building, Trumpington Street, Cambridge CB2 1RP
32 East 57th Street, New York, NY 10022, USA
296 Beaconsfield Parade, Middle Park, Melbourne 3206, Australia

First published 1981

Printed in Great Britain at the
University Press, Cambridge

British Library cataloguing in publication data
School Mathematics Project
Individualised mathematics.
Probability and statistics
1. Mathematics – 1961–
I. Title   II. National Extension College
510   OA39.2   80–49964
ISBN 0 521 23367 4

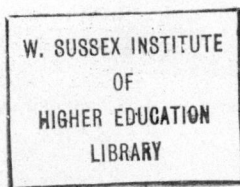